E
SOURCE
The Prentice Hall
ENGINEERING SOURCE

Introduction to AutoCAD® R. 14

Mark Dix and Paul Riley

CAD Support Associates

Prentice Hall
Upper Saddle River, NJ 07458

Library of Congress Cataloging-In-Publication Data

Dix, Mark
Introduction to AutoCAD R. 14 / Mark Dix and Paul Riley
p. cm.
Includes bibliographical references and index.
ISBN 0-13-011001-9
1. Computer graphics. 2. AutoCAD (Computer file). I. Riley, Paul, 1943– . II. Title.
T385.D6327 1998
604.2 '0285'5369—dc21 97-50975
CIP

Editor-in-chief: **MARCIA HORTON**
Acquisitions editor: **ERIC SVENDSEN**
Director of production and manufacturing: **DAVID W. RICCARDI**
Managing editor: **EILEEN CLARK**
Editorial/production supervision: **ROSE KERNAN**
Cover director: **JAYNE CONTE**
Creative director: **AMY ROSEN**
Manufacturing buyer: **PAT BROWN**
Editorial assistant: **GRIFFIN CABLE**
Marketing manager: **DANNY HOYT**

Printed in the United States of America

10 9 8 7 6 5 4 3 2 1

ISBN 0-13-011001-9

Prentice-Hall International (UK) Limited, *London*
Prentice-Hall of Australia Pty. Limited, *Sydney*
Prentice-Hall Canada, Inc., *Toronto*
Prentice-Hall Hispanoamericana, S.A., *Mexico*
Prentice-Hall of India Private Limited, *New Delhi*
Prentice-Hall of Japan, Inc., *Tokyo*
Simon & Schuster Asia Pte., Ltd., *Singapore*
Editora Prentice-Hall do Brazil, Ltda., *Rio de Janeiro*

Drawing credits: Hearth Drawing courtesy of Thomas Casey: Double Bearing Assembly compliments of David Sumner, King Philip Technical Drawing; Isometric Drawing Flanged Coupling courtesy of Richard F. Ross; Isometric Drawing of Garage courtesy of Thomas Casey.

About ESource

The Challenge

Professors who teach the Introductory/First-Year Engineering course popular at most engineering schools have a unique challenge—teaching a course defined by a changing curriculum. The first-year engineering course is different from any other engineering course in that there is no real cannon that defines the course content. It is not like Engineering Mechanics or Circuit Theory where a consistent set of topics define the course. Instead, the introductory engineering course is most often defined by the creativity of professors and students, and the specific needs of a college or university each semester. Faculty involved in this course typically put extra effort into it, and it shows in the uniqueness of each course at each school.

Choosing a textbook can be a challenge for unique courses. Most freshmen require some sort of reference material to help them through their first semesters as a college student. But because faculty put such a strong mark on their course, they often have a difficult time finding the right mix of materials for their course and often have to go without a text, or with one that does not really fit. Conventional textbooks are far too static for the typical specialization of the first-year course. How do you find the perfect text for your course that will support your students educational needs, but give you the flexibility to maximize the potential of your course?

ESource—The Prentice Hall Engineering Source http://emissary.prenhall.com/esource

Prentice Hall created ESource—The Prentice-Hall Engineering Source—to give professors the power to harness the full potential of their text and their freshman/first year engineering course. In today's technologically advanced world, why settle for a book that isn't perfect for your course? Why not have a book that has the exact blend of topics that you want to cover with your students?

More then just a collection of books, ESource is a unique publishing system revolving around the ESource website—http://emissary.prenhall.com/esource/. ESource enables you to put your stamp on your book just as you do your course. It lets you:

Control You choose exactly what chapters or sections are in your book and in what order they appear. Of course, you can choose the entire book if you'd like and stay with the authors original order.

Optimize Get the most from your book and your course. ESource lets you produce the optimal text for your students needs.

Customize You can add your own material anywhere in your text's presentation, and your final product will arrive at your bookstore as a professionally formatted text.

ESource Content

All the content in ESource was written by educators specifically for freshman/first-year students. Authors tried to strike a balanced level of presentation, one that was not either too formulaic and trivial, but not focusing heavily on advanced topics that most introductory students will not encounter until later classes. A developmental editor reviewed the books and made sure that every text was written at the appropriate level, and that the books featured a balanced presentation. Because many professors do not have extensive time to cover these topics in the classroom, authors prepared each text with the idea that many students would use it for self-instruction and independent study. Students should be able to use this content to learn the software tool or subject on their own.

While authors had the freedom to write texts in a style appropriate to their particular subject, all followed certain guidelines created to promote the consistency a text needs. Namely, every chapter opens with a clear set of objectives to lead students into the chapter. Each chapter also contains practice problems that tests a student's skill at performing the tasks they have just learned. Chapters close with extra practice questions and a list of key terms for reference. Authors tried to focus on motivating applications that demonstrate how engineers work in the real world, and included these applications throughout the text in various chapter openers, examples, and problem material. Specific Engineering and Science **Application Boxes** are also located throughout the texts, and focus on a specific application and demonstrating its solution.

Because students often have an adjustment from high school to college, each book contains several **Professional Success Boxes** specifically designed to provide advice on college study skills. Each author has worked to provide students with tips and techniques that help a student better understand the material, and avoid common pitfalls or problems first-year students often have. In addition, this series contains an entire book titled ***Engineering Success*** by Peter Schiavone of the University of Alberta intended to expose students quickly to what it takes to be an engineering student.

Creating Your Book

Using ESource is simple. You preview the content either on-line or through examination copies of the books you can request on-line, from your PH sales rep, or by calling(1-800-526-0485). Create an on-line outline of the content you want in the order you want using ESource's simple interface. Either type or cut and paste your own material and insert it into the text flow. You can preview the overall organization of the text you've created at anytime (please note, since this preview is immediate, it comes unformatted.), then press another button and receive an order number for your own custom book . If you are not ready to order, do nothing—ESource will save your work. You can come back at any time and change, re-arrange, or add more material to your creation. You are in control. Once you're finished and you have an ISBN, give it to your bookstore and your book will arrive on their shelves six weeks after the order. Your custom desk copies with their instructor supplements will arrive at your address at the same time.

To learn more about this new system for creating the perfect textbook, go to **http://emissary.prenhall.com/esource/**. You can either go through the on-line walkthrough of how to create a book, or experiment yourself.

Community

ESource has two other areas designed to promote the exchange of information among the introductory engineering community, the Faculty and the Student Centers. Created and maintained with the help of Dale Calkins, an Associate Professor at the University of Washington, these areas contain a wealth of useful information and tools. You can preview outlines created by other schools and can see how others organize their courses. Read a monthly article discussing important topics in the curriculum. You can post your own material and share it with others, as well as use what others have posted in your own documents. Communicate with our authors about their books and make suggestions for improvement. Comment about your course and ask for information from others professors. Create an on-line syllabus using our custom syllabus builder. Browse Prentice Hall's catalog and order titles from your sales rep. Tell us new features that we need to add to the site to make it more useful.

Supplements

Adopters of ESource receive an instructor's CD that includes solutions as well as professor and student code for all the books in the series. This CD also contains approximately **350 Powerpoint Transparencies** created by Jack Leifer—of University South Carolina—Aiken. Professors can either follow these transparencies as pre-prepared lectures or use them as the basis for their own custom presentations. In addition, look to the web site to find materials from other schools that you can download and use in your own course.

Titles in the ESource Series

Introduction to Unix
0-13-095135-8
David L. Schwartz

Introduction to Maple
0-13-095-133-1
David L. Schwartz

Introduction to Word
0-13-254764-3
David C. Kuncicky

Introduction to Excel
0-13-254749-X
David C. Kuncicky

Introduction to MathCAD
0-13-937493-0
Ronald W. Larsen

Introduction to AutoCAD, R. 14
0-13-011001-9
Mark Dix and Paul Riley

Introduction to the Internet, 2/e
0-13-011037-X
Scott D. James

Design Concepts for Engineers
0-13-081369-9
Mark N. Horenstein

Engineering Design—A Day in the Life of Four Engineers
0-13-660242-8
Mark N. Horenstein

Engineering Ethics
0-13-784224-4
Charles B. Fleddermann

Engineering Success
0-13-080859-8
Peter Schiavone

Mathematics Review
0-13-011501-0
Peter Schiavone

Introduction to ANSI C
0-13-011854-0
Dolores Etter

Introduction to C++
0-13-011855-9
Dolores Etter

Introduction to MATLAB
0-13-013149-0
Dolores Etter

Introduction to FORTRAN 90
0-13-013146-6
Larry Nyhoff & Sanford Leestma

http://emissary.prenhall.com/esource/

About the Authors

No project could ever come to pass without a group of authors who have the vision and the courage to turn a stack of blank paper into a book. The authors in this series worked diligently to produce their books, provide the building blocks of the series.

Delores M. Etter is a Professor of Electrical and Computer Engineering at the University of Colorado. Dr. Etter was a faculty member at the University of New Mexico and also a Visiting Professor at Stanford University. Dr. Etter was responsible for the Freshman Engineering Program at the University of New Mexico and is active in the Integrated Teaching Laboratory at the University of Colorado. She was elected a Fellow of the Institute of Electrical and Electronic Engineers for her contributions to education and for her technical leadership in digital signal processing. IN addition to writing best-selling textbooks for engineering computing, Dr. Etter has also published research in the area of adaptive signal processing.

Sanford Leestma is a Professor of Mathematics and Computer Science at Calvin College, and received his Ph.D from New Mexico State University. He has been the long time co-author of successful textbooks on Fortran, Pascal, and data structures in Pascal. His current research interests are in the areas of algorithms and numerical compuitation.

Larry Nyhoff is a Professor of Mathematics and Computer Science at Calvin College. After doing bachelors work at Calvin, and Masters work at Michigan, he received a Ph.D. from Michigan State and also did graduate work in computer science at Western Michigan. Dr. Nyhoff has taught at Calvin for the past 34 years—mathematics at first and computer science for the past several years. He has co-authored several computer science textbooks since 1981 including titles on Fortran and C++, as well as a brand new title on Data Structures in C++.

Acknowledgments: We express our sincere appreciation to all who helped in the preparation of this module, especially our acquisitions editor Alan Apt, managing editor Laura Steele, development editor Sandra Chavez, and production editor Judy Winthrop. We also thank Larry Genalo for several examples and exercises and Erin Fulp for the Internet address application in Chapter 10. We appreciate the insightful review provided by Bart Childs. We thank our families—Shar, Jeff, Dawn, Rebecca, Megan, Sara, Greg, Julie, Joshua, Derek, Tom, Joan; Marge, Michelle, Sandy, Lori, Michael—for being patient and understanding. We thank God for allowing us to write this text.

Mark Dix began working with AutoCAD in 1985 as a programmer for CAD Support Associates, Inc. He helped design a system for creating estimates and bills of material directly from AutoCAD drawing databases for use in the automated conveyor industry. This system became the basis for systems still widely in use today. In 1986 he began collaborating with Paul Riley to create AutoCAD training materials, combining Riley's background in industrial design and training with Dix' s background in writing, curriculum development, and programming. Dix and Riley have created tutorial and teaching methods for every AutoCAD release since Version 2.5. Mr. Dix has a Master of Arts in Teaching from Cornell University and a Masters of Education from the University of Massachusetts. He is currently the Director of Dearborn Academy High School in Arlington, Massachusetts.

Paul Riley is an author, instructor, and designer specializing in graphics and design for multimedia. He is a founding partner of CAD Support Associates, a contract service and professional training organization for computer-aided design. His 15 years of business experience and 20 years of teaching experience are supported by degrees

in education and computer science. Paul has taught AutoCAD at the University of Massachusetts at Lowell and is presently teaching AutoCAD at Mt. Ida College in Newton, Massachusetts. He has developed a program, Computer-Aided Design for Professionals that is highly regarded by corporate clients and has been an ongoing success since 1982.

David I. Schwartz is a Lecturer at SUNY-Buffalo who teaches freshman and first-year engineering, and has a Ph.D from SUNY-Buffalo in Civil Engineering. Schwartz originally became interested in Civil engineering out of an interest in building grand structures, but has also pursued other academic interests including artificial intelligence and applied mathematics. He became interested in Unix and Maple through their application to his research, and eventually jumped at the chance to teach these subjects to students. He tries to teach his students to become incremental learners and encourages frequent practice to master a subject, and gain the maturity and confidence to tackle other subjects independently. In his spare time, Schwartz is an avid musician and plays drums in a variety of bands.

Acknowledgments: I would like to thank the entire School of Engineering and Applied Science at the State University of New York at Buffalo for the opportunity to teach not only my students, but myself as well; all my EAS140 students, without whom this book would not be possible—thanks for slugging through my lab packets; Andrea Au, Eric Svendsen, and Elizabeth Wood at Prentice Hall for advising and encouraging me as well as wading through my blizzard of e-mail; Linda and Tony for starting the whole thing in the first place; Rogil Camama, Linda Chattin, Stuart Chen, Jeffrey Chottiner, Roger Christian, Anthony Dalessio, Eugene DeMaitre, Dawn Halvorsen, Thomas Hill, Michael Lamanna, Nate "X" Patwardhan, Durvejai Sheobaran, "Able" Alan Somlo, Ben Stein, Craig Sutton, Barbara Umiker, and Chester "JC" Zeshonski for making this book a reality; Ewa Arrasjid, "Corky" Brunskill, Bob Meyer, and Dave Yearke at "the Department Formerly Known as ECS" for all their friendship, advice, and respect; Jeff, Tony, Forrest, and Mike for the interviews; and, Michael Ryan and Warren Thomas for believing in me.

Ronald W. Larsen is an Associate Professor in Chemical Engineering at Montana State University, and received his Ph.D from the Pennsylvania State University. Larsen was initially attracted to engineering because he felt it was a serving profession, and because engineers are often called on to eliminate dull and routine tasks. He also enjoys the fact that engineering rewards creativity and presents constant challenges. Larsen feels that teaching large sections of students is one of the most challenging tasks he has ever encountered because it enhances the importance of effective communication. He has drawn on a two year experince teaching courses in Mongolia through an interpreter to improve his skills in the classroom. Larsen sees software as one of the changes that has the potential to radically alter the way engineers work, and his book Introduction to Mathcad was written to help young engineers prepare to be productive in an ever-changing workplace.

Acknowledgments: To my students at Montana State University who have endured the rough drafts and typos, and who still allow me to experiment with their classes—my sincere thanks.

Peter Schiavone is a professor and student advisor in the Department of Mechanical Engineering at the University of Alberta. He received his Ph.D. from the University of Strathclyde, U.K. in 1988. He has authored several books in the area of study skills and academic success as well as numerous papers in scientific research journals.

Before starting his career in academia, Dr. Schiavone worked in the private sector for Smith's Industries (Aerospace and Defence Systems Company) and Marconi Instruments in several different areas of engineering including aerospace, systems and software engineering. During that time he developed an interest

http://emissary.prenhall.com/esource/

in engineering research and the applications of mathematics and the physical sciences to solving real-world engineering problems.

His love for teaching brought him to the academic world. He founded the first Mathematics Resource Center at the University of Alberta: a unit designed specifically to teach high school students the necessary survival skills in mathematics and the physical sciences required for first-year engineering. This led to the Students' Union Gold Key award for outstanding contributions to the University and to the community at large.

Dr. Schiavone lectures regularly to freshman engineering students, high school teachers, and new professors on all aspects of engineering success, in particular, maximizing students' academic performance. He wrote the book *Engineering Success* in order to share with you the *secrets of success in engineering study*: the most effective, tried and tested methods used by the most successful engineering students.

Acknowledgments: I'd like to acknowledge the contributions of: Eric Svendsen, for his encouragement and support; Richard Felder for being such an inspiration; the many students who shared their experiences of first-year engineering—both good and bad; and finally, my wife Linda for her continued support and for giving me Conan.

Scott D. James is a staff lecturer at Kettering University (formerly GMI Engineering & Management Institute) in Flint, Michigan. He is currently pursuing a Ph.D. in Systems Engineering with an emphasis on software engineering and computer-integrated manufacturing. Scott decided on writing textbooks after he found a void in the books that were available. "I really wanted a book that showed how to do things in good detail but in a clear and concise way. Many of the books on the market are full of fluff and force you to dig out the really important facts." Scott decided on teaching as a profession after several years in the computer industry. "I thought that it was really important to know what it was like outside of academia. I wanted to provide students with classes that were up to date and provide the information that is really used and needed."

Acknowledgments: Scott would like to acknowledge his family for the time to work on the text and his students and peers at Kettering who offered helpful critique of the materials that eventually became the book.

David C. Kuncicky is a native Floridian. He earned his Baccalaureate in psychology, Master's in computer science, and Ph.D. in computer science from Florida State University. He is also the author of *Excel 97 for Engineers.* Dr. Kuncicky is the Director of Computing and Multimedia Services for the FAMU-FSU College of Engineering. He also serves as a faculty member in the Department of Electrical Engineering. He has taught computer science and computer engineering courses for the past 15 years. He has published research in the areas of intelligent hybrid systems and neural networks. He is actively involved in the education of computer and network system administrators and is a leader in the area of technology-based curriculum delivery.

Acknowledgments: Thanks to Steffie and Helen for putting up with my late nights and long weekends at the computer. Thanks also to the helpful and insightful technical reviews by the following people: Jerry Ralya, Kathy Kitto of Western Washington University, Avi Singhal of Arizona State University, and Thomas Hill of the State University of New York at Buffalo. I appreciate the patience of Eric Svendsen and Rose Kernan of Prentice Hall for gently guiding me through this project. Finally, thanks to Dean C.J. Chen for providing continued tutelage and support.

Mark Horenstein is an Associate Professor in the Electrical and Computer Engineering Department at Boston University. He received his Bachelors in Electrical Engineering in 1973 from Massachusetts Institute of Technology, his Masters in Electrical Engineering in 1975

from University of California at Berkeley, and his Ph.D. in Electrical Engineering in 1978 from Massachusetts Institute of Technology. Professor Horenstein's research interests are in applied electrostatics and electromagnetics as well as microelectronics, including sensors, instrumentation, and measurement. His research deals with the simulation, test, and measurement of electromagnetic fields. Some topics include electrostatics in manufacturing processes, electrostatic instrumentation, EOS/ESD control, and electromagnetic wave propagation.

Professor Horenstein designed and developed a class at Boston University, which he now teaches entitled Senior Design Project (ENG SC 466). In this course, the student gets real engineering design experience by working for a virtual company, created by Professor Horenstein, that does real projects for outside companies—almost like an apprenticeship. Once in "the company" (Xebec Technologies), the student is assigned to an engineering team of 3-4 persons. A series of potential customers are recruited, from which the team must accept an engineering project. The team must develop a working prototype deliverable engineering system that serves the need of the customer. More than one team may be assigned to the same project, in which case there is competition for the customer's business.

Acknowledgements: Several individuals contributed to the ideas and concepts presented in Design Principles for Engineers. The concept of the Peak Performance design competition, which forms a cornerstone of the book, originated with Professor James Bethune of Boston University. Professor Bethune has been instrumental in conceiving of and running Peak Performance each year and has been the inspiration behind many of the design concepts associated with it. He also provided helpful information on dimensions and tolerance. Several of the ideas presented in the book, particularly the topics on brainstorming and teamwork, were gleaned from a workshop on engineering design help bi-annually by Professor Charles Lovas of Southern Methodist University. The principles of estimation were derived in part from a freshman engineering problem posed by Professor Thomas Kincaid of Boston University.

I would like to thank my family, Roxanne, Rachel, and Arielle, for giving me the time and space to think about and write this book. I also appreciate Roxanne's inspiration and help in identifying examples of human/machine interfaces.

Dedicated to Roxanne, Rachel, and Arielle

Charles B. Fleddermann is a professor in the Department of Electrical and Computer Engineering at the University of New Mexico in Albuquerque, New Mexico. He is a third generation engineer—his grandfather was a civil engineer and father an aeronautical engineer—so "engineering was in my genetic makeup." The genesis of a book on engineering ethics was in the ABET requirement to incorporate ethics topics into the undergraduate engineering curriculum. "Our department decided to have a one-hour seminar course on engineering ethics, but there was no book suitable for such a course." Other texts were tried the first few times the course was offered, but none of them presented ethical theory, analysis, and problem solving in a readily accessible way. "I wanted to have a text which would be concise, yet would give the student the tools required to solve the ethical problems that they might encounter in their professional lives."

http://emissary.prenhall.com/esource/

Reviewers

ESource benefited from a wealth of reviewers who on the series from its initial idea stage to its completion. Reviewers read manuscripts and contributed insightful comments that helped the authors write great books. We would like to thank everyone who helped us with this project.

Concept Document
Naeem Abdurrahman- University of Texas, Austin
Grant Baker- University of Alaska, Anchorage
Betty Barr- University of Houston
William Beckwith- Clemson University
Ramzi Bualuan- University of Notre Dame
Dale Calkins- University of Washington
Arthur Clausing- University of Illinois at Urbana-Champaign
John Glover- University of Houston
A.S. Hodel- Auburn University
Denise Jackson- University of Tennessee, Knoxville
Kathleen Kitto- Western Washington University
Terry Kohutek- Texas A&M University
Larry Richards- University of Virginia
Avi Singhal- Arizona State University
Joseph Wujek- University of California, Berkeley
Mandochehr Zoghi- University of Dayton

Books
Stephen Allan- Utah State University
Naeem Abdurrahman - University of Texas Austin
Anil Bajaj- Purdue University
Grant Baker - University of Alaska - Anchorage
Betty Barr - University of Houston
William Beckwith - Clemson University
Haym Benaroya- Rutgers University
Tom Bledsaw- ITT Technical Institute
Tom Bryson- University of Missouri, Rolla
Ramzi Bualuan - University of Notre Dame
Dan Budny- Purdue University
Dale Calkins - University of Washington
Arthur Clausing - University of Illinois
James Devine- University of South Florida
Patrick Fitzhorn - Colorado State University
Dale Elifrits- University of Missouri, Rolla
Frank Gerlitz - Washtenaw College
John Glover - University of Houston
John Graham - University of North Carolina-Charlotte
Malcom Heimer - Florida International University
A.S. Hodel - Auburn University
Vern Johnson- University of Arizona
Kathleen Kitto - Western Washington University
Robert Montgomery- Purdue University
Mark Nagurka- Marquette University
Ramarathnam Narasimhan- University of Miami
Larry Richards - University of Virginia
Marc H. Richman - Brown University
Avi Singhal-Arizona State University
Tim Sykes- Houston Community College
Thomas Hill- SUNY at Buffalo
Michael S. Wells - Tennessee Tech University
Joseph Wujek - University of California - Berkeley
Edward Young- University of South Carolina
Mandochehr Zoghi - University of Dayton

Contents

4 Template Drawings

5 Arcs and Polar Arrays

6 Object Snap

7 Text

8 Dimensions

1

Lines

OVERVIEW

This chapter will introduce you to some of the tools you will use whenever you draw in AutoCAD. You will begin to find your way around the Release 14 for menus and toolbars and you will learn to control basic elements of the Drawing Window. You will produce drawings involving straight lines and learn to undo your last command with the U command. Your drawings will be saved, if you wish, using the SAVE or SAVEAS commands.

SECTIONS

OBJECTIVES

After reading this chapter, you should be able to:

- Begin a new drawing.
- Explore the drawing window.
- Explore comands entry methods.
- Draw and undo lines.
- Save and open drawings.
- Review material.
- Draw a grate.
- Draw a design.
- Draw a shim.

1.1 BEGINNING A NEW DRAWING

When you load Release 14 you will find yourself in the Drawing Window with the Startup dialog box open. The Drawing Window is where you do most of your work with AutoCAD. You can begin a new drawing or open a previously saved file. In this task you will begin a new drawing and ensure that your Drawing Window shows the "Start from Scratch" default settings we have used in preparing this chapter.

⌖ **From the Windows 95 startup screen, choose Programs, then the AutoCAD Release 14 folder and then the AutoCAD Release 14 program.**

Or, if your system has an AutoCAD Release 14 shortcut icon on the Windows 95 desktop, double click the icon to start AutoCAD.

⌖ **Wait . . . .**

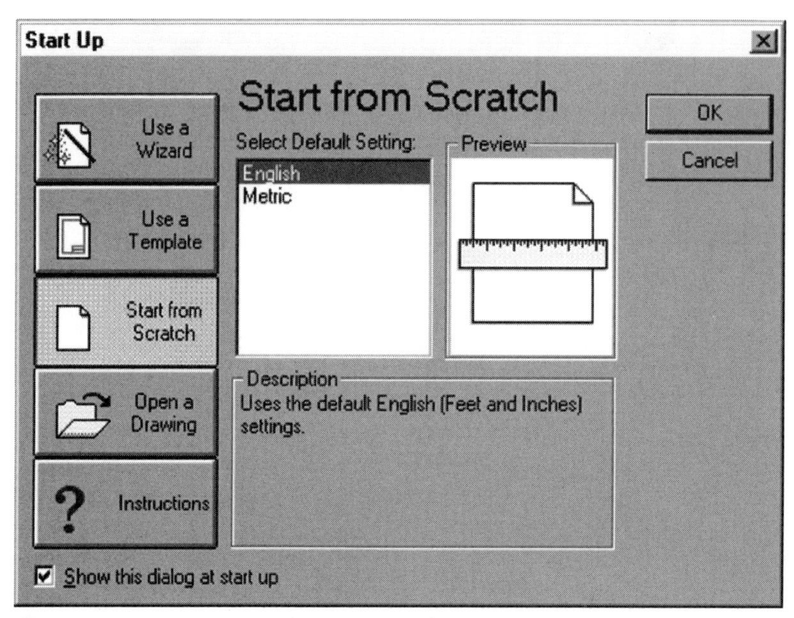

Figure 1.1.

When you see the Release 14 screen with the Startup dialog box in the foreground as shown in *Figure 1.1* you are ready to begin.

The Startup dialog box gives you several options including the use of a Wizard, which guides you through setup, a template, which uses a previously saved template drawing as the basis for your new drawing, or starting from scratch. For our purposes it will be best to start from scratch to ensure that your Drawing Window shows the same settings as are used in this chapter. We will explore other possibilities later.

You should see your cursor arrow somewhere on the screen. If you do not see it, move your pointing device to the middle of your drawing area or digitizer. When you see the arrow, you are ready to proceed.

Click on "Start from Scratch."

Click on "English" in the Select Default Setting box.

Click on Okay to exit the dialog and start your new drawing. Pressing Enter will also work.

The dialog box will disappear and your screen should resemble *Figure 1.2*. You are now ready to proceed.

1.2 EXPLORING THE DRAWING WINDOW

You are looking at the AutoCAD Drawing Window. There are many ways that you can alter it to suit a particular drawing application. To begin with, there are a number of features that can be turned on and off using your mouse or the F-keys on your keyboard. Note that the six function keys and their functions are listed at the bottom of every drawing through Chapter 8 of this book, so it is not necessary to memorize them now. You will learn them best through repeated use.

The Screen

The Release 14 Drawing Window has many features that are common to all Windows 95 programs. At the top of the screen you will see the title bar, with the AutoCAD

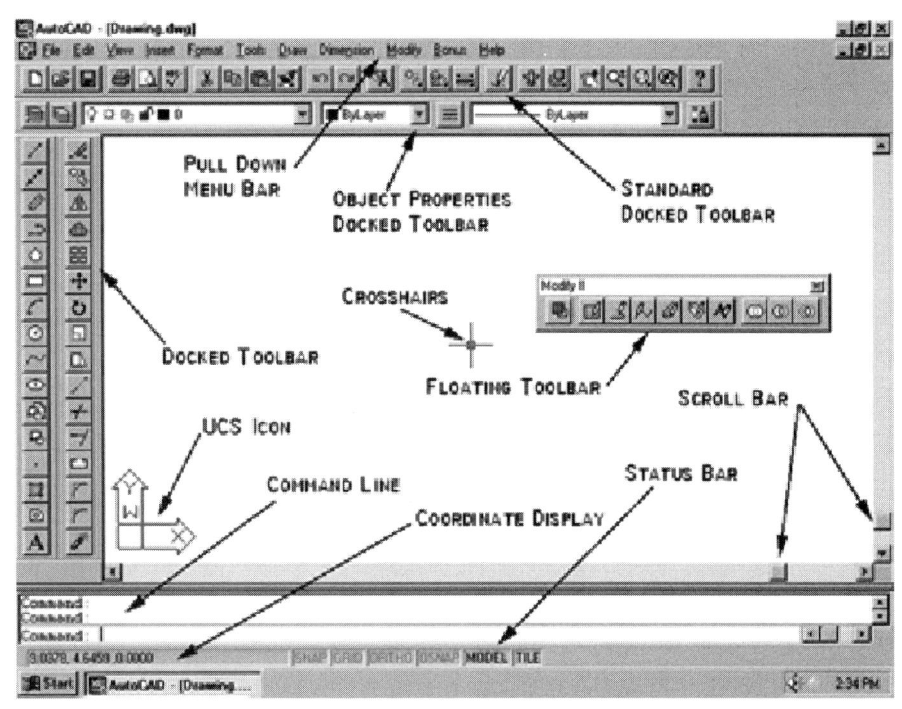

Figure 1.2.

Release 14 icon on the left and the standard Windows 95 minimize, maximize, and close buttons on the right. Next to the word "AutoCAD" you will also see "[Drawing]." This is AutoCAD's default drawing name. When you have named and saved your drawing, the drawing name will appear in the brackets. This drawing name is also displayed at the bottom of the screen on the Windows 95 taskbar.

Below the title bar you will see Release 14's pull down menu bar including the titles for the File, Edit, View, Insert, Format, Tools, Draw, Dimension, Modify, and Help pull down menus. Pull down menus are discussed later in this task.

The next line down is the standard toolbar. This toolbar is one of many toolbars that can be displayed on the Release 14 screen. The use of toolbars will also be discussed later in this chapter and in Chapter 3.

The next line is the Object Properties toolbar which displays the current layer and linetype. It also includes tools for changing other object properties. Layers and linetypes are discussed in Chapter 4.

Below these toolbars you will see the drawing area on the right with the Draw and Modify toolbars positioned vertically on the left.

Below the drawing area and along the right side of the screen you will see scroll bars. These work like the scroll bars in any Windows application. Clicking on the arrows or clicking and dragging the square sliders will move your drawing to the left or right, up or down within the drawing area. You should have no immediate need for this function.

Beneath the scroll bar you will see the command prompt area. Typed commands are one of the basic ways of working in AutoCAD.

At the bottom of the Drawing Window is the status bar, with the coordinate display on the left and six mode indicators (Snap, Grid, Ortho, Osnap, Model and Tile) in the middle. The coordinate display and mode indicators are discussed shortly.

Finally, the bottom of your screen will show the Windows 95 taskbar, with the Start button on the left and any open applications following. You should see the Release 14 icon and the name of your drawing here, as at the top of your screen.

TIP: You can expand your drawing window slightly if you wish by hiding the Windows 95 taskbar. This can be done by pointing to the top of the taskbar and dragging it down off the bottom of the screen. It will move down leaving only the top edge showing. To get it back, point to this top edge again and drag it upwards.

Switching Screens

⌖ **Press F2.**

This brings up the AutoCAD text window. AutoCAD uses this window to display text that will not fit in the command area. As soon as the text window opens, Windows 95 adds a button for it on the taskbar at the bottom of the screen. Now you can open and close the text window using the taskbar buttons or F2.

⌖ **Press F2 again, or click on the AutoCAD—[Drawing] button on the taskbar.**

This brings you back to the Drawing Window. Once used, the AutoCAD Text Window operates like an open application. It remains "open" but may be visible or invisible until you close it using the close button in the upper right corner.

⌖ **Press F2 or the AutoCAD text window button to view the text window again.**

⌖ **Click on the close button (X) in the upper right corner of the text window to close the text window.**

NOTE: Sometimes AutoCAD switches to the text window automatically when there is not enough room in the command area for prompts or messages. If this happens, use F2 when you are ready to return to the drawing area.

The Mouse

Your pointing device has many functions and will handle most of your interaction with AutoCAD. Given the toolbar and menu structure of AutoCAD and Windows 95, a two button mouse is sufficient for most applications. In this book we will assume two buttons. If you have a digitizer and a more complex pointing device the two button functions will be present along with other functions that we will not address.

On a common two button mouse the left button will be used for point selection, object selection, and menu or tool selection. All mouse instructions, refer to this left button, unless specifically stated otherwise. The right button functions as an alternative to the enter key on the keyboard, and when used in the right context, calls pop up menus or help windows. Learning how and when to use these right button functions can immediately increase your efficiency. We will point them out as we go along.

Cross Hairs and Pickbox

You should see small cross with a box at its intersection somewhere in the display area of your screen. If you do not see it, move your pointing device until it appears. The two perpendicular lines are the cross hairs, or screen cursor, that tell you where your pointing device is located on your digitizer or mouse pad.

The small box at the intersection of the cross hairs is called the "pick box" and is used to select objects for editing. You will learn more about the pickbox later.

Move the pointer and see how the cross hairs move in coordination with your hand movements.

⌗ **Move the pointer so that the cross hairs move to the top of the screen.**

When you leave the drawing area, your cross hairs will be left behind and you will see an arrow pointing up and to the left. The arrow will be used as in other Windows 95 applications to select tools and to pull down menus from the menu bar.

NOTE: Here and throughout this book we show the Release 14 Windows 95 versions of AutoCAD screens in our illustrations. If you are working with another version, your screen may show significant variations.

⌗ **Move the cursor back into the drawing area and the selection arrow will disappear.**

Toolbars and Pull Down Menus

There are seventeen toolbars available in the standard AutoCAD Release 14 toolbar dialog box. Toolbars can be created and modified. They can be moved, resized, and reshaped. They are a convenience, but they can also make your work and your drawing area overly cluttered. For our purposes you will not need more than a few of the available toolbars.

The Coordinate Display

The coordinate display at the bottom left of the status line keeps track of coordinates as you move the pointer. The coordinate display is controlled using the F6 key.

Move the cross hairs around slowly and keep your eye on the three numbers at the bottom of the screen. They should be moving very rapidly through four-place decimal numbers. When you stop moving, the numbers will be showing coordinates for the location of the pointer. These coordinates are standard coordinate values in a three dimensional coordinate system originating from (0,0,0) at the lower left corner. The first value is the x value, showing the horizontal position of the cross hairs, measuring left to right. The second value is y, or the vertical position of the cross hairs, measured from bottom to top. Points also have a z value, but it will always be 0 in two-dimensional drawing and can be ignored until you begin to draw in 3D (Chapter 9). *In this book we will usually not include the z value if it is 0, as it will be until we get into 3D drawings.*

⌗ **Press F6.**

The numbers will freeze and the coordinate display will turn gray.

⌗ **Move the cross hairs slowly.**

Now when you move the cross hairs you will see that the coordinate display does not change. You will also see that it is grayed out. The display is now in static mode in which the numbers will only change when you select a new point by pressing the pick button on your cursor. Try it.

⌗ **Move the cross hairs to any point on the screen and press the left button on your mouse. The left button is the pick button.**

Notice that the first two numbers in the coordinate display change, but the display is still "grayed out."

⌗ **Move the cross hairs to another point on the screen.**

You will see that AutoCAD opens a box on the screen, as shown in *Figure 1.3.* This is the **object selection window.** It will have no effect now since there are no objects on your screen. Object selection is discussed Chapter 2.

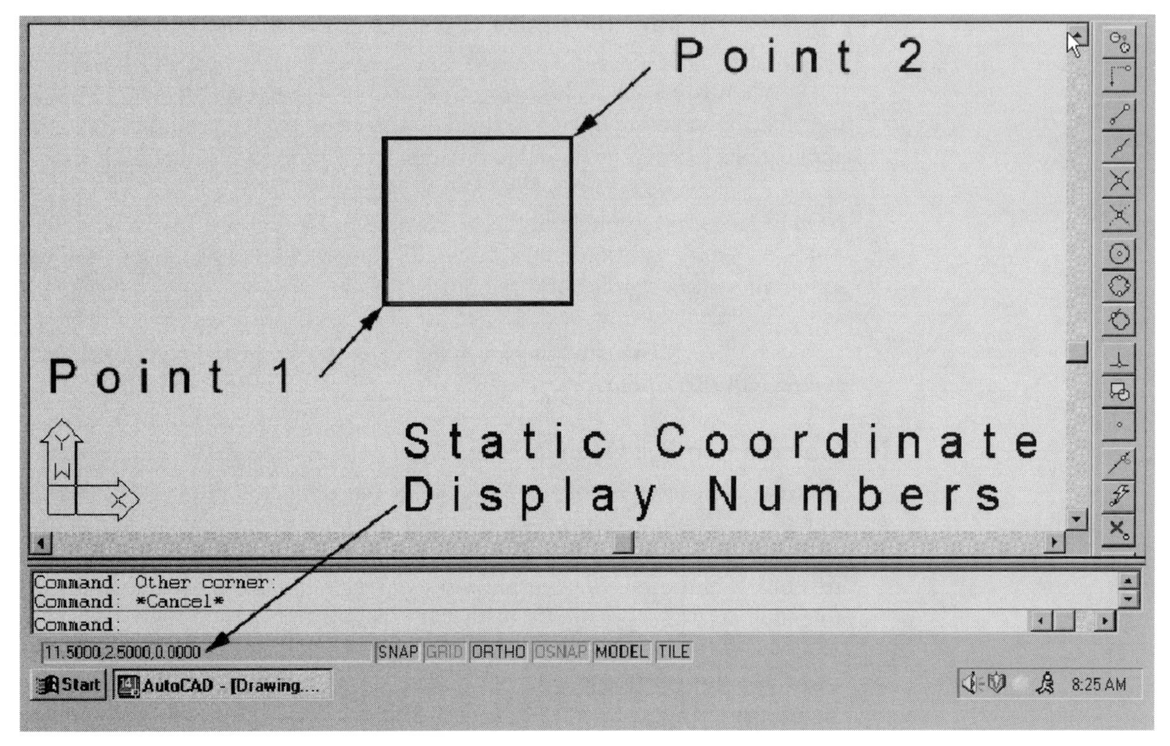

Figure 1.3.

AutoCAD prompts for the other corner of the box. You will see this in the command area:

Other corner:

Pick a second point.

This will complete the object selection window. Notice the change in the static coordinate display numbers

Press F6 to return the coordinate display to dynamic mode.

The coordinate display actually has two different dynamic modes, but this will not be apparent until you enter a drawing command such as LINE (Task 3) which asks for point selection.

NOTE: The units AutoCAD uses for coordinates, dimensions, and for measuring distances and angles can be changed at any time using the UNITS command (Chapter 2). For now we will accept the AutoCAD default values, including the four-place decimals. In the next chapter we will be changing to two-place decimals. *The F-keys are switches only, they cannot be used to change settings.*

The Grid

Press F7 or double click on the word GRID on the status bar.

This will turn on the grid. When the grid is on, the word "GRID" will become black on the status bar.

The grid is simply a matrix of dots that helps you find your way around on the screen. It will not appear on your drawing when it is plotted, and it may be turned on and off at will. You may also change the spacing between dots, using the GRID command as we will be doing in Chapter 2.

The grid is presently set up to emulate the shape of an A-size (9 × 12 inch) piece of paper. There are 10 grid points from bottom to top, numbered 0 to 9, and 13 points from left to right, numbered 0 to 12. The AutoCAD command that controls the outer size and shape of the grid is LIMITS, which will be discussed in Chapter 4. Until then we will continue to use the present format.

Also, you should be aware from the beginning that there is no need to scale AutoCAD drawings while you are working on them. That can be handled when you get ready to plot or print your drawing on paper. You will always draw at full scale, where one unit of length on the screen equals one unit of length in real space. This full scale drawing space is called "model space." Notice the word "MODEL" on the status bar, indicating that you are currently working in model space. The actual size of drawings printed out on paper may be handled in "paper space." When you are in paper space you will see the word "PAPER" on the status bar in place of the word "MODEL." For now, all your work will be done in model space and you do not need to be concerned with paper space. The word "TILE" at the right of the status bar also affects model space and paper space, but is of no concern to you at this point.

Snap

⌖ **Press F9 or double click on the word "SNAP" on the status bar.**

SNAP should become black indicating that snap mode is now on.

⌖ **Move the cross hairs slowly around the drawing area.**

Look closely and you will see how the cross hairs jump from point to point. If your grid is on, notice that it is impossible to make the cross hairs touch a point that is not on the grid. Try it.

You will also see that the coordinate display shows only values ending in .0000 or .5000.

⌖ **Press F9 or double click on SNAP again.**

Snap should now be off and SNAP will be grayed out on the status bar again.

If you move the cursor in a circle now, you will see that the cross hairs move more smoothly, without jumping. You will also observe that the coordinate display moves rapidly through a full range of four-place decimal values again.

F9 turns snap on and off. With snap off you can, theoretically, touch every point on the screen. With snap on you can move only in predetermined increments. With the Start from Scratch default settings snap is set to a value of 0.5000 so that you can move only in half unit increments. In the next chapter you will learn how to change this setting using the SNAP command. For now we will leave the snap setting alone. A snap setting of 0.5000 will be adequate for the drawings at the end of this chapter.

Using an appropriate snap increment is a tremendous timesaver. It allows for a degree of accuracy that is not possible otherwise. If all the dimensions in a drawing fall into one-inch increments, for example, there is no reason to deal with points that are not on a one-inch grid. You can find the points you want much more quickly and accurately if all those in between are temporarily eliminated. The snap setting will allow you to do that.

Ortho

F8 turns ortho mode on and off. You will not observe the ortho mode in action until you have entered the LINE command, however. We will try it out at the end of Task 3.

The User Coordinate System Icon

At the lower left of the screen you will see the User Coordinate System (UCS) icon (see *Figure 1.2*). These two perpendicular arrows clearly indicate the directions of the X and Y axes, which are currently aligned with the sides of your screen. In Chapter 12, when you begin to do 3D drawings, you will be defining your own coordinate systems that can be turned at any angle and originate at any point in space. At that time you will find that the icon is a very useful visual aid. However, it is hardly necessary in two-dimensional drawing and may be distracting. For this reason you may want to turn it off now and keep it turned off until you actually need it.

⌖ **Type "ucsicon."**

Notice that the typed letters are displayed on the command line to the right of the colon.

⌖ **Press the enter key on your keyboard.**

AutoCAD will show the following prompt on the command line:

ON/OFF/All/Noorigin/ORigin <ON>:

As you explore AutoCAD commands you will become familiar with many prompts like this one. It is simply a series of options separated by slashes (/). For now we only need to know about "On" and "Off."

⌖ **Type "off" and press enter.**

The UCS icon will disappear from your screen. Anytime you want to see it again, type "ucsicon" and then type "on." Alternatively, you can click on "View" on the menu bar, then highlight "Display" and "Ucsicon" and click "On." This will turn the icon on if it is off and off if it is on.

1.3 EXPLORING COMMAND ENTRY METHODS

You can communicate drawing instructions to AutoCAD by typing or by selecting items from a toolbar, a pull down menu, or a tablet menu. Each method has its advantages and disadvantages depending on the situation. Often a combination of two or more methods is the most efficient way to carry out a complete command sequence. The instructions in this book are not always specific about which to use. All operators develop their own preferences.

Each method is described briefly. You do not have to try them all out at this time. Read them over to get a feel for the possibilities and then proceed to the LINE command. As a rule, we suggest learning the keyboard procedure first. It is the most basic, the most comprehensive, and changes the least from one release of AutoCAD to the next. But do not limit yourself by typing everything. As soon as you know the keyboard sequence, try out the other methods to see how they vary and how you can use them to save time. Ultimately, you will want to type as little as possible and use the differences between the menu systems to your advantage.

The Keyboard and the Command Line

The keyboard is the most primitive and fundamental method of interacting with AutoCAD. Toolbars, pull down menus, and tablet menus all function by automating basic command sequences as they would be typed on the keyboard. It is therefore useful to be familiar with the keyboard procedures even if the other methods are sometimes faster.

As you type commands and responses to prompts, the characters you are typing will appear on the command line after the colon. Remember that you must press enter to complete your commands and responses. The command line can be moved and reshaped, or you can switch to the text screen using F2 when you want to see more lines including previously typed entries.

It is very useful to know that some of the most often used commands, such as LINE, ERASE, and CIRCLE have "aliases." These one- or two-letter abbreviations are very handy. A few of the most commonly used aliases are shown in *Figure 1.4.* In Release 14 there are a large number of two-letter aliases, which we will introduce as we go along.

Pull Down Menus

Pull down menus and toolbars have the advantage that instead of typing a complete command, you can simply "point and shoot" to select an item. In Release 14, the pull down menus contain most commands that you will use regularly. Menu selections and toolbar selections often duplicate each other.

Pull down menus work here as they do in any Windows 95 application. To use the a menu, move the cross hairs up into the menu bar so that the selection arrow appears. Then move it to the menu heading you want. Select it with the pick button (the left button on your mouse). A menu will appear. Run down the list of items to the one you want. Press the pick button again to select the item (see *Figure 1.5*). Items followed by a triangle have "cascading" submenus. Submenus open automatically when an item is highlighted. Picking an item that is followed by an ellipsis (. . .) will call up a dialog box.

Dialog boxes are familiar features in many Windows and Macintosh programs. They require a combination of pointing and typing that is fairly intuitive. We will discuss many dialog boxes in detail as we go along.

Toolbars

Toolbars are also a standard Windows feature. They are comprised of buttons with icons that give one click access to commands. Seventeen toolbars can be opened from the Toolbars dialog box at the bottom of the View pull down menu.

Once opened, toolbars can "float" anywhere on the screen, or can be "docked" along the edges of the drawing area. Toolbars can be a nuisance, since they cover portions of your drawing space, but they can be opened and closed quickly. Beyond the Standard, Object Properties, Draw, and Modify toobars which are open by default, you

Figure 1.4.

COMMAND ALIAS CHART		
LETTER	+ ENTER	= COMMAND
A	↵	ARC
C	↵	CIRCLE
E	↵	ERASE
L	↵	LINE
M	↵	MOVE
P	↵	PAN
R	↵	REDRAW
Z	↵	ZOOM

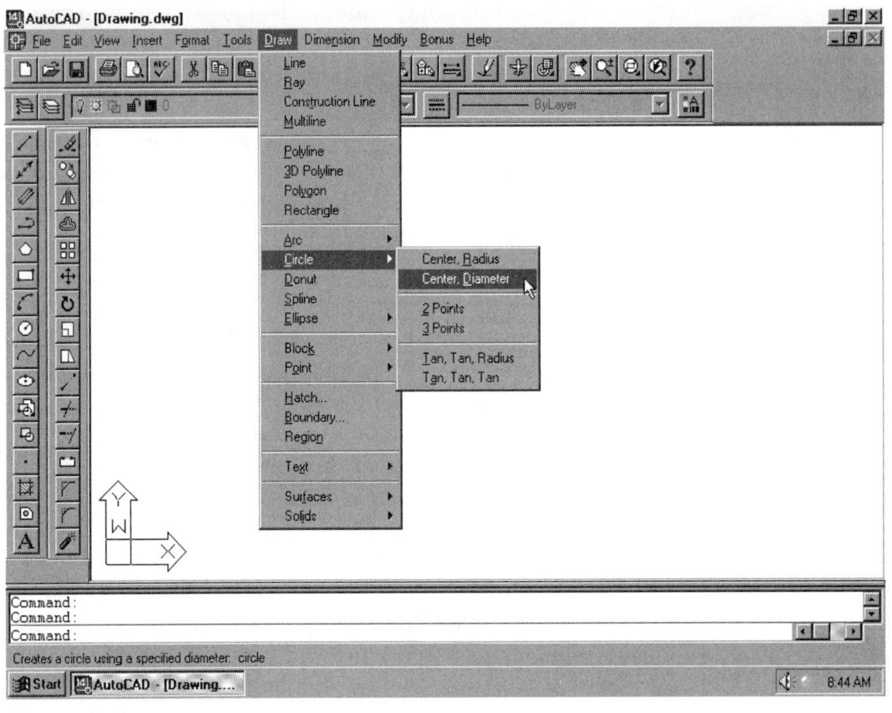

Figure 1.5.

will probably only wish to open a toolbar if you are doing a whole set of procedures involving one toolbar. In dimensioning an object, for example, you may wish to have the Dimensioing toolbar open. Do not use too many toolbars at once and remember that you can move toolbars or use the scroll bars to move your drawing right, left, up, and down behind the toolbars.

The icons used on toolbars are also a mixed blessing. One picture may be worth a thousand words, but with so many pictures, you may find that a few words can be very helpful as well. As in other Windows applications, you can get a label for an icon simply by allowing the selection arrow to rest on the button for a moment without selecting it. These labels are called "Tooltips." Try this:

⌖ **Position the selection arrow on the top button of the Draw toolbar, as shown in *Figure 1.6*, but do not press the pick button.**

⌖ **You will see a yellow label that says "Line", as shown in the figure. This label identifies this button as the LINE command button.**

When a tooltip is displayed, you will also see a phrase in the status bar in place of the coordinate display. This phrase will describe what the tool or menu item does and is called a "Helpstring." The LINE helpstring says "Creates straight line segments: line." The word following the colon identifies the command as you would type it in the command area.

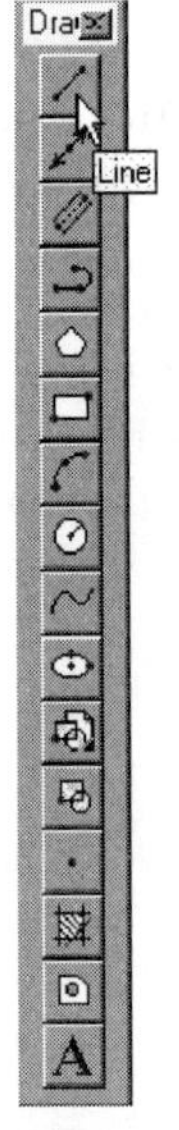

Figure 1.6.

1.4 DRAWING AND UNDOING LINES

Now it is time to enter the LINE command and start drawing.

⌖ **Type "L" or select the Line icon from the Draw toolbar, or Line under Draw on the pull down menu (remember to press enter if you are typing).**

Look at the command area. You should see this, regardless of how you enter the command:

From point:

This is AutoCAD's way of asking for a start point.

Also, notice that the pickbox disappears from the cross hairs when you have entered a drawing command.

Most of the time when you are drawing you will want to point rather than type. In order to do this you need to pay attention to the grid and the coordinate display.

⌗ If snap is off switch it on (F9 or double click on SNAP).

⌗ Move the cursor until the display reads "1.0000,1.0000, 0.000." Then press the pick button.

Now AutoCAD will ask for a second point. You should see this in the command area:

To point:

Rubber Band

There are two new things to be aware of. One is the "rubber band" that extends from the start point to the cross hairs on the screen. If you move the cursor, you will see that this visual aid stretches, shrinks, or rotates like a tether keeping you connected to the start point. You will also notice that when the rubber band and the cross hairs overlap (i.e., when the rubber band is at 0, 90, 180, or 270 degrees) they both disappear in the area between the cross hairs and the start point, as illustrated in *Figure 1.7*. .This may seem odd at first, but it is actually a great convenience. You will find many instances where you will need to know that the cross hairs and the rubber band are exactly lined up

Absolute (XYZ) and Polar Coordinates

The other thing to watch is the coordinate display. If it is off (no change in coordinates when the cursor moves), press F6 to turn it on. Once it is on it will show either absolute coordinates or polar coordinates. Absolute coordinates are the familiar coordinate numbers discussed previously. They are distances measured from the origin (0,0,0) of the

Figure 1.7.

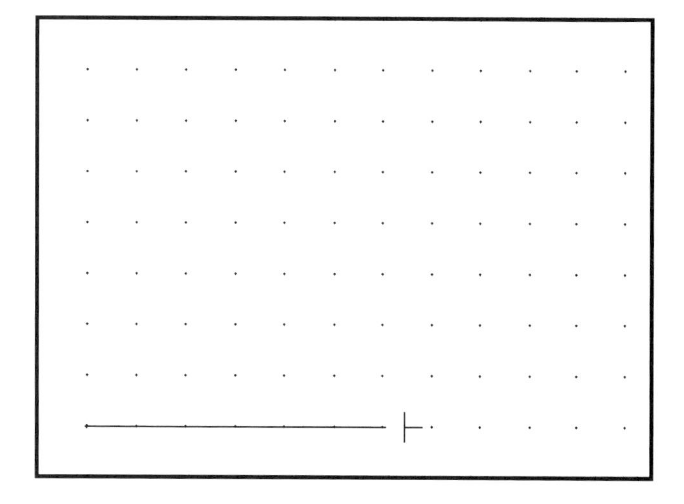

coordinate system at the lower left corner of the grid. If your display shows three four-digit numbers, then these are the absolute coordinates.

If your display shows something like “4.2426 < 45,0.0000”, it is set on polar coordinates.

Press F6 and move your cursor.

Which type of coordinates is displayed?

Press F6 and move your cursor again.

Observe the coordinate display.

You will see that there are three coordinate display modes: static (no change until you select a point), *xyz* (*x*, *y*, and *z* values separated by a comma), and polar (length < angle, z).

Press F6 once or twice until it shows polar coordinates.

Polar coordinates are in a length, angle, z format and are given relative to the last point you picked. They look something like this: 5.6569 < 45,0.0000. There are three values here: 5.6569, 45, and 0.0000. The first number (5.6569) is the distance from the starting point of the line to the cross hairs. The second (45) is an angle of rotation, measuring counter-clockwise with 0 degrees being straight out to the right. The third value (0.0000) is the z coordinate which will remain zero in 2D drawing. In LINE, as well as most other draw commands, polar coordinates are very useful because the first number will give you the length of the segment you are currently drawing.

Press F6 to read xyz coordinates.

Pick the point (8.0000,8.0000, 0.0000).

Your screen should now resemble *Figure 1.8.* AutoCAD has drawn a line between (1,1) and (8,8) and is asking for another point.

To point:

This will allow you to stay in the LINE command to draw a whole series of connected lines if you wish. You can draw a single line from point to point, or a series of lines from point to point to point to point. In either case, you must tell AutoCAD when you are finished by pressing enter or the enter equivalent button on the cursor, or the space bar.

Figure 1.8.

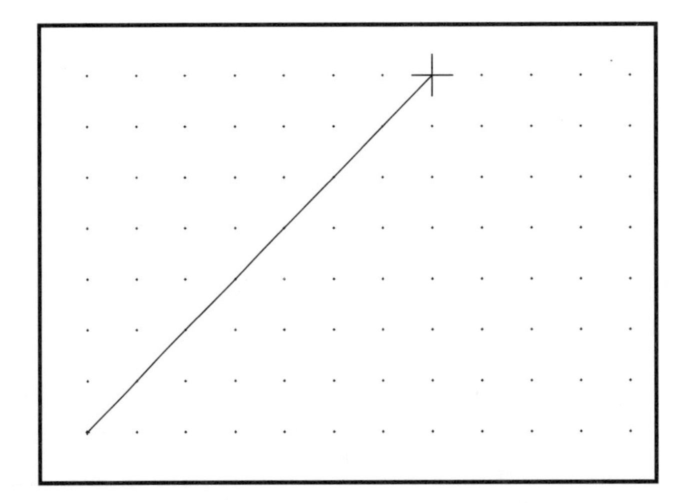

NOTE: When you are drawing a continuous series of lines, the polar coordinates on the display are given relative to the most recent point, not the original starting point.

⌖ **Press enter or the space bar to end the LINE command.**

You should be back to the "Command:" prompt again, and the pickbox will have reappeared at the intersection of the cross hairs.

Space Bar, Enter Key, and Enter Button

In most cases AutoCAD allows you to use the space bar as a substitute for the enter key or the enter button on your mouse. This is often a convenience, since the space bar is easy to locate with one hand while the other hand is on the pointing device. For example, the LINE command can be entered without removing your right hand from your pointing device by typing the "L" on the keyboard and then hitting the space bar, both with the left hand. The major exception to the use of the space bar as an enter key is when you are entering text in the TEXT, DTEXT, or MTEXT commands (Chapter 7). Since a space may be part of a text string, the space bar must have its usual significance there.

TIP: Hitting the space bar, the enter key, or the enter button on your pointing device at the command prompt will repeat the last command entered. This isa major convenience, and we recommend that you use it often.

Direct Distance Entry

Another convenient drawing method, new in Release 14, is called Direct Distance Entry. In this method, you pick the first point, then show the direction of the line segment you wish to draw, but instead of picking the other end point you type in a value for the length of the line. Try it:

Repeat the LINE command by pressing enter, the space bar, or the enter (right) button on your mouse.

AutoCAD prompts for a first point.

TIP: If you press enter, the space bar, or the enter button at the "From point:" prompt, AutoCAD will select the last point entered, so that you can begin drawing from where you left off.

Press enter, the space bar, or the enter button on your mouse to select the point (8,8,0), the end point of the previously drawn line.

AutoCAD prompts for a second point.

⌖ **Pull the rubber band diagonally down to the right, as shown in Figure 1.9.**

Moving along the diagonal set of grid points will put you at a –45 degree angle as shown. The length of the rubber band does not matter, only the direction.

⌖ **With the rubber band stretched out as shown, type "3."**

⌖ **Press enter, or hit the space bar.**

AutoCAD will draw a 3.0000 line segment at the angle you have shown.

⌖ **Press enter, the enter button on your pointing device, or the space bar to exit the LINE command.**

Your screen should resemble *Figure 1.10.*

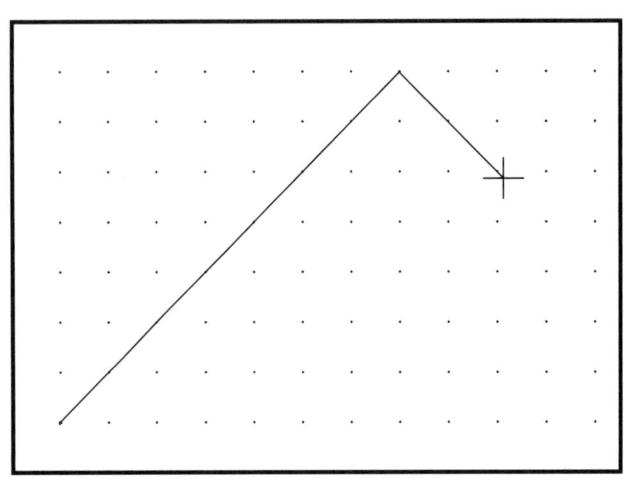

Figure 1.9.

Undoing Commands with U

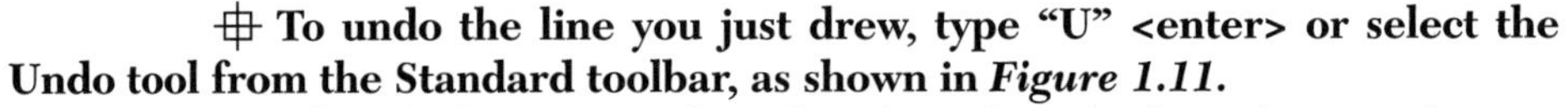

⌗ **To undo the line you just drew, type "U" <enter> or select the Undo tool from the Standard toolbar, as shown in *Figure 1.11.***

Figure 1.11.

U undoes the last command, so if you have done anything else since drawing the line, you will need to type "U" <enter> more than once. In this way you can walk backwards through your drawing session undoing your commands one by one.

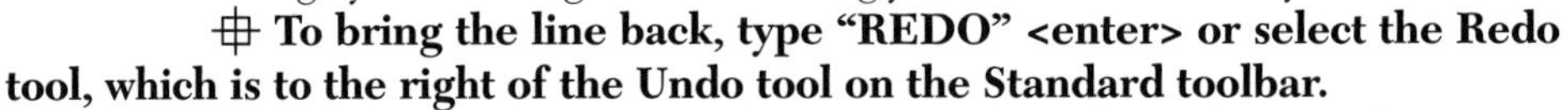

⌗ **To bring the line back, type "REDO" <enter> or select the Redo tool, which is to the right of the Undo tool on the Standard toolbar.**

REDO only works immediately after U, and it only works once. That is, you can only REDO the last U and only if it was the last command executed.

⌗ **Before going on, enter U twice, or as many times as necessary to undo any lines you have drawn and leave your drawing area blank.**

Ortho

Before completing this section, we suggest that you try the ortho mode.

Figure 1.10.

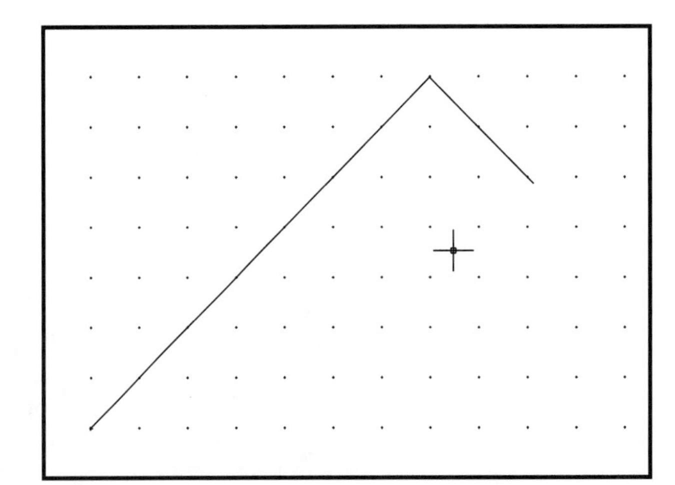

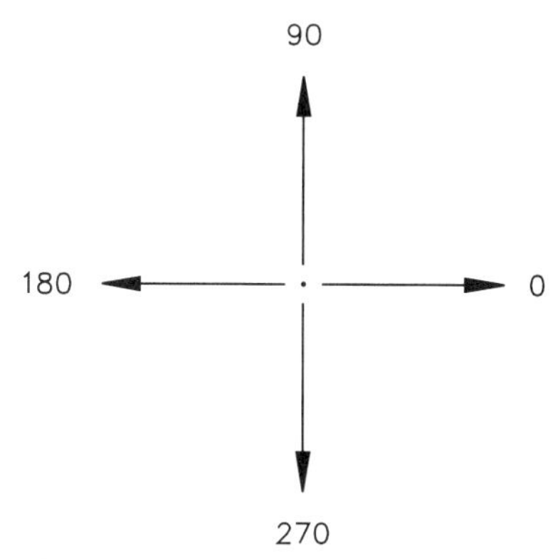

Figure 1.12.

⌗ Type "L" <enter> or select Line or the Line tool.
⌗ Pick a starting point. Any point near the center of the screen will do.
⌗ Press F8 or double click on the word ORTHO and move the cursor in slow circles.

Notice how the rubber band jumps between horizontal and vertical without sweeping through any of the angles between. Ortho forces the pointing device to pick up points only along the horizontal and vertical quadrant lines from a given starting point. With ortho on you can select points at 0, 90, 180, and 270 degrees of rotation from your starting point only (see *Figure 1.12*).

The advantages of ortho are similar to the advantages of snap mode, except that it limits angular rather than linear increments. It ensures that you will get precise and true right angles and perpendiculars easily when that is your intent. Ortho will become more important as drawings grow more complex. In this chapter it is hardly necessary, though it will be convenient in drawings 1 and 3.

The Esc key

⌗ While still in the LINE command, press the "Esc" (escape) key.

This will abort the LINE command and bring back the "Command": prompt. Esc is used to cancel a command that has been entered. Sometimes it will be necessary to press Esc twice to exit a command and return to the command prompt.

1.5 SAVING AND OPENING YOUR DRAWINGS

Saving drawings in Release 14 is just like saving a file in most other Windows 95 applications. Use SAVE to save an already named drawing. Use SAVEAS to name a drawing as you save it, or save a named drawing under a new name. In all cases, a .dwg extension is added to file names to identify them as AutoCAD files. This is automatic when you name a file.

Figure 1.13.

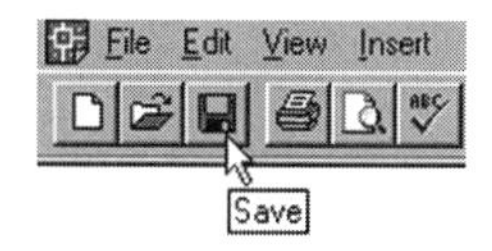

The SAVE Command

To save your drawing without leaving the drawing window, select "Save" from the File pull down menu, or select the Save tool from the Standard toolbar, as shown in *Figure 1.13*.

If the current drawing is already named, Release 14 will save it without intervening dialog. If it is not named, it will open the SAVEAS dialog box and allow you to give it a name before it is saved. SAVEAS is described following.

The SAVEAS Command

To rename a drawing or to give a name to a previously unnamed drawing type "Saveas", select "Saveas . . ." from the File pull down menu, or, in an unnamed drawing, select the Save tool from the Standard Toolbar.

Any of these methods will call the Save Drawing As dialog box (see *Figure 1.14*). The cursor will be blinking in the area labelled "Files:", waiting for you to enter a file name. Include a drive designation, (i.e., "A:1-1") if you are saving your work on a floppy disk. AutoCAD will add the .dwg extension automatically. SAVEAS will also allow you to save different versions of the same drawing under different names while continuing to edit.

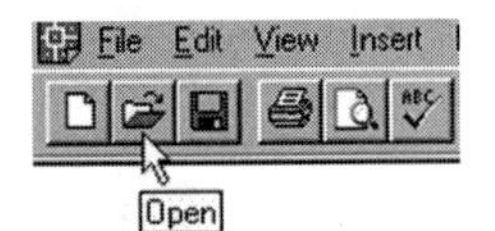

Figure 1.15.

OPENing Saved Drawings

To open a previously saved drawing, type "open", select "Open" from the File pull down menu, or select the Open tool from the Standard toolbar, as shown in *Figure 1.15.*

Once you have saved a drawing you will need to use the OPEN command to return to it later. Entering the OPEN command by any method will bring up the Select File dialog box shown in *Figure 1.16.* You can select a file directory in the box at the left and then a file from the box at the right. When you select a file, Release 14 will show a preview image of the selected drawing in the Preview image box at the right. This way you can make sure that you are opening the drawing you want.

Figure 1.14.

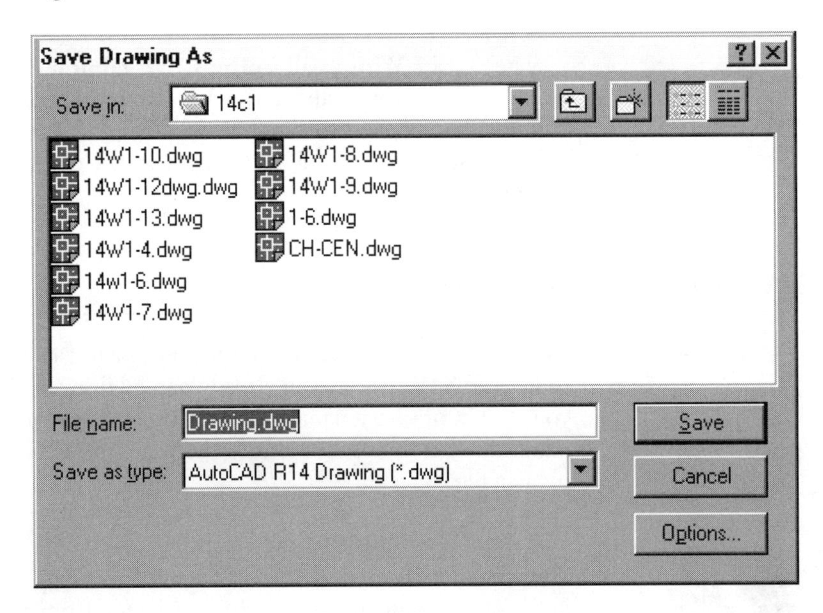

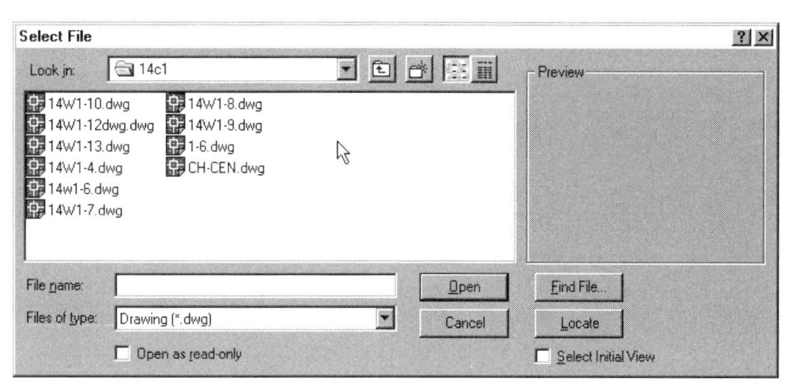

Figure 1.16.

1.6 REVIEW MATERIAL

Questions

Before going to the drawings, quickly review the following questions and problems. Then you should be ready for Drawing #1.

1. What function key opens and closes the text window?
2. What are the three different modes of the coordinate display and how does each mode appear? How do you switch between modes?
3. Explain and describe the differences among absolute, relative, and polar coordinates.
4. You have just entered the point (1,1,0) and you now wish to enter the point 2 units straight up from this point. How would you identify this point using absolute, relative, and polar coordinates.
5. What is the value and limitation of having SNAP on?
6. Name three different ways to enter the LINE command.
7. Name and describe three different methods of point selection in AutoCAD.
8. What does the U command do?
9. What is the main limitation of the REDO command?
10. What key do you use to cancel a command?
11. What command would you use to save a new version of a drawing under a new file name? How would you enter it?

COMMANDS

Draw	Edit	File
LINE	REDO	NEW
	U	SAVE
		SAVEAS
		OPEN

View
UCSICON

Drawing Problems

1. Draw a line from (3,2) to (4,8) using the keyboard only.
2. Draw a line from (6,6) to (7,5) using the mouse only.
3. Draw a line from (6,6) to (6,8) using the direct distance method.
4. Undo (U) all lines on your screen.
5. Draw a square with corners at (2,2), (7,2), (7,7), and (2,7). Then erase it using the U command as many times as necessary.

TIP: Closing a Set of Lines
For drawing an enclosed figure like the one in problem 5, the LINE command provides a convenient Close option. "Close" will connect the last in a continuous series of lines back to the starting point of the series. In drawing a square, for instance, you would simply type "C" <enter> in lieu of drawing the last of the four lines. In order for this to work, the whole square must be drawn without leaving the LINE command.

PROFESSIONAL SUCCESS

Will the Computer Replace Pencil and Paper Drawings?

Not too long ago, a student entering an engineering graphics class would likely encounter rows of large drafting tables used in the production of precision drawings by hand. Today, computerized drawing programs like AutoCAD have, for the most part, replaced mechanical drawing.

Though computer aided drafting software has become almost ubiquitous, technical sketching by hand is still an essential tool for the successful engineer. Just as an electronic calculator is not used in every circumstance (some calculations are more easily done in your head), hand sketching sometimes proves to be the most convenient means of rapidly producing a drawing.

Hand sketches, with practice, should be neither imprecise nor sloppy; any type of engineering drawing that can be produced via computer or mechanical drafting methods can be rapidly realized by sketch.

A major advantage of sketching is that it can be done anywhere. Quick sketches may be produced by an engineer on a plant tour for the purpose of recording a particular design or layout. A group of engineers at a restaurant or party may resort to sketching on napkins as they work as a group to refine an idea into a conceptual design. In such cases, sketches are being used as a means of communications among engineers, as it is often easier to express an idea pictorially than in words. Until computers are as small, light and as easy to use as a pencil and tablet (or a napkin!) hand sketching will remain among the most important skills that an engineer can master.

DRAWING 1—1

Grate

Before beginning, look over the drawing page. Notice the F-key reminders and other drawing information at the bottom. The commands on the right are the new commands you will need to do this drawing. They are listed with their toolbar or menu locations in parentheses.

The first two drawings in this chapter are given without dimensions. Instead, we have drawn them as you will see them on the screen, against the background of a half-unit grid. All of these drawings were done using a half-unit snap, and all points will be found on one-unit increments.

Drawing Suggestions

- If you are beginning a new drawing, type or select "New . . .?" and then select Start from Scratch in the Startup dialog box. Notice that Release 14 remembers your last selection and it becomes the default. Because of this you may be able to simply press enter or select okay to Startup.
- Remember to watch the coordinate display when searching for a point.
- Be sure that grid, snap, and the coordinate display are all turned on.
- Draw the outer rectangle first. It is 6 units wide and 7 units high, and its lower left-hand corner is at the point (3.0000,1.0000). The three smaller rectangles inside are 4×1.
- The Close option can be used in all four of the rectangles.

If You Make A Mistake - U The U command works nicely within the LINE command to undo the last line you drew, or the last two or three if you have drawn a series.

- Type "U" <enter>. The last line you drew will be gone or replaced by the rubber band awaiting a new end point. If you want to go back more than one line, type "U" <enter> again, as many times as you need.
- If you have already left the LINE command, U will still work, but instead of undoing the last line, it will undo the last continuous series of lines. In Grate this could be the whole outside rectangle, for instance.
- Remember, if you have mistakenly undone something, you can get it back by typing "REDO" <enter>. You cannot perform other commands between U and REDO.

U is quick, easy to use, and efficient as long as you always spot your mistakes immediately after making them. Most of us, however, are more spontaneous in our blundering. We may make mistakes at any time and not notice them until the middle of next week. For us, AutoCAD provides more flexible editing tools, like ERASE, which is introduced in the next chapter.

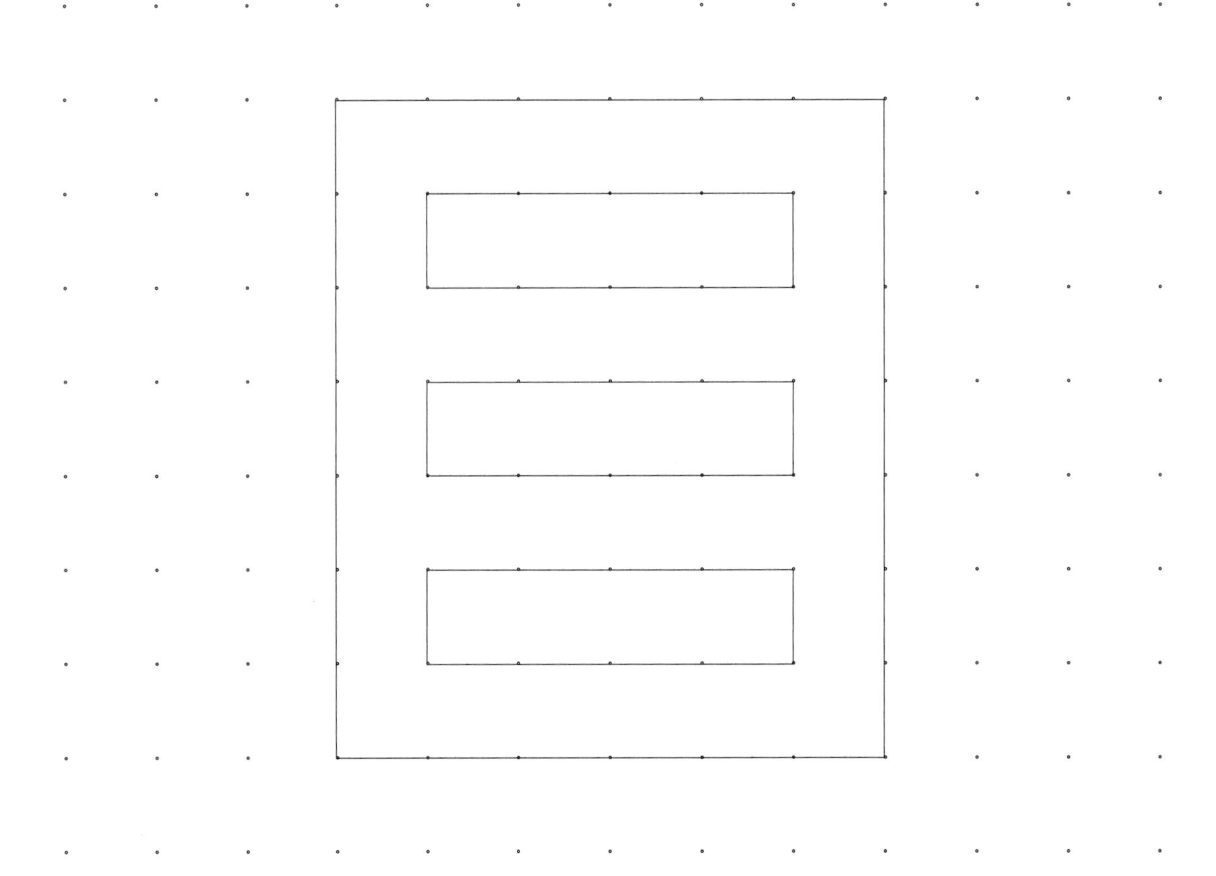
GRATE
Drawing 1-1

F1
HELP
F2
TEXT/GRAPHICS SCREEN
F6
ABSOLUTE/OFF/POLAR COORDS
F7
ON/OFF GRID
F8
ON/OFF ORTHO
F9
ON/OFF SNAP

DRAWING 1—2

Design

This design will give you further practice with the LINE command.

Drawing Suggestions

- If you are beginning a new drawing, type or select "New . . .?" and then check the Start Up dialog box to ensure that you Start from Scratch.
- Draw the horizontal and vertical lines first. Each is eight units long.
- Notice how the rest of the lines work—outside point on horizontal to inside point on vertical, then working in, or vice-versa.

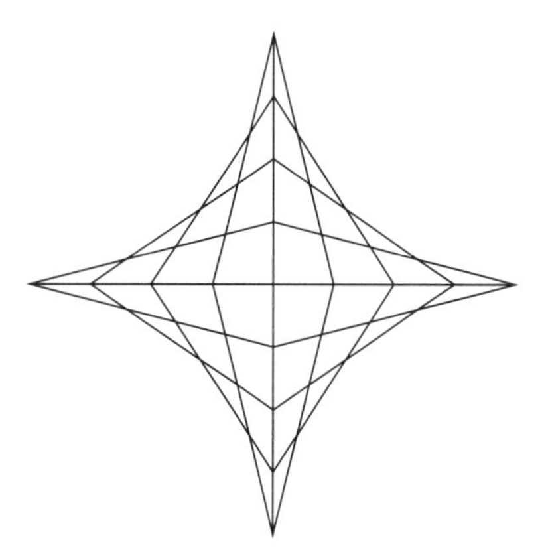

Repeating a Command Remember, you can repeat a command by pressing enter, the enter button on your cursor, or the space bar at the Command: prompt. This will be useful in this drawing, since you have several sets of lines to draw.

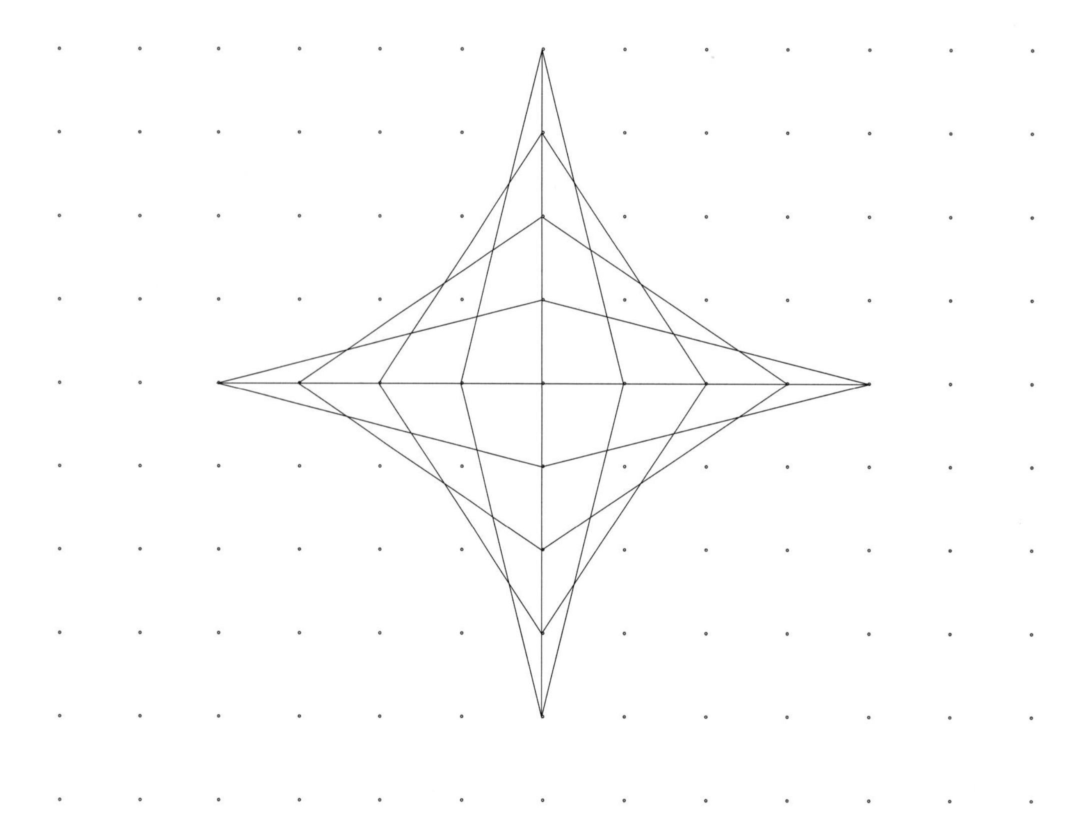

DESIGN # 1

Drawing 1–2

F1	F2	F6	F7	F8	F9
HELP	TEXT/GRAPHICS SCREEN	ABSOLUTE/OFF/POLAR COORDS	ON/OFF GRID	ON/OFF ORTHO	ON/OFF SNAP

DRAWING 1—3

Shim

This drawing will give you further practice in using the LINE command. In addition, it will give you practice in translating dimensions into distances on the screen. Note that the dimensions are only included for your information; they are not part of the drawing at this point. Your drawing will appear like the reference drawing that follows. Dimensioning is discussed in Chapter 8.

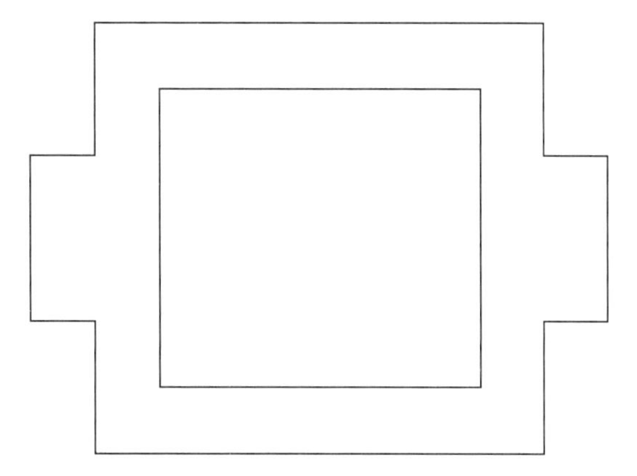

Drawing Suggestions

- If you are beginning a new drawing, type or select "New . . .?" and Start from Scratch.
- It is most important that you choose a starting point that will position the drawing so that it fits on your screen. If you begin with the bottom left-hand corner of the outside figure at the point (3,1), you should have no trouble.
- Read the dimensions carefully to see how the geometry of the drawing works. It is good practice to look over the dimensions before you begin drawing. Often the dimension for a particular line may be located on another side of the figure or may have to be extrapolated from other dimensions. It is not uncommon to misread, misinterpret, or miscalculate a dimension, so take your time.

9.0000
7.0000
1.0000
1.0000
3.0000
7.0000
2.0000
1.0000
2.0000
5.0000
SQUARE

SHIM
Drawing 1–3

F1	F2	F6	F7	F8	F9
HELP	TEXT/GRAPHICS SCREEN	ABSOLUTE/OFF/POLAR COORDS	ON/OFF GRID	ON/OFF ORTHO	ON/OFF SNAP

2

Circles and Drawing Aids

OVERVIEW

In this chapter you will learn to change the spacing of the grid and the snap.You will also change the units in which coordinates are displayed.You will explore the CIRCLE command and AutoCAD's HELP features, and learn to delete objects selectively using ERASE.You will also learn to measure and mark distances on the screen using the DIST command. Finally, you begin to learn AutoCAD plotting and printing procedures.

SECTIONS

OBJECTIVES

After reading this chapter, you should be able to:

- Change the grid spacing.
- Change the snap spacing.
- Change units.
- Draw three concentric circles using the center point, radius method.
- Draw three more concentric circles using the center point, diameter method.
- Access AutoCAD HELP features.
- ERASE the circles, using four selection methods.
- Print or plot a drawing.
- Draw an aperture wheel.
- Draw a roller.
- Draw a switch plate.

2.1 CHANGING THE GRID SETTING

When you begin a new drawing from scratch in Release 14, the grid and snap are set with a spacing of 0.5000. In Chapter 1 all drawings were done without altering the grid and snap spacings from the default value. Frequently you will want to change this, depending on your application. You may want a 10-foot snap for a building layout, or a 0.010 inch snap for a printed circuit diagram. The grid may match the snap setting, or may be set independently.

To begin, open a new drawing by typing "new" or selecting New from the File menu or the Standard toolbar, and then selecting Start from Scratch and English units in the Start up dialog box.

Once again, this ensures that you begin with the settings we have used in preparing this chapter.

Using F7 and F9, or double clicking in the status bar, be sure that GRID and SNAP are both on.

⌗ **Type "grid"** (we will no longer remind you to press enter after typing a command or response to a prompt). Do not use the pull down menu or toolbar yet. We will get to those procedures momentarily.

The command area prompt will appear like this, with options separated by slashes (/):

Grid spacing(X) or ON/OFF/Snap/Aspect <0.5000>:

You can ignore the options for now. The number <0.5000> shows the present setting. AutoCAD uses this format (<default>) in many command sequences to show you a current value or default setting. It usually comes at the end of a series of options. Pressing the enter key or space bar at this point will confirm the default setting.

⌗ **In answer to the prompt, type "1" and watch what happens** (of course you remembered to press enter).

The screen will change to show a one unit grid.

⌗ **Move the cursor around to observe the effects of the new grid setting.**

The snap setting has not changed, so you still have access to all half unit points, but the grid shows only single unit increments.

⌗ **Try other grid settings. Try 2, .25, and .125.**

Remember that you can repeat the last command, GRID, by pressing enter, the space bar, or the enter key on your pointing device.

⌗ **What happens when you try .05?**

When you get too small (smaller than .106 on our screen), the grid becomes too dense to display, but the snap can still be set smaller.

Using The Drawing Aids Dialog Box

You can also change grid and snap settings using a dialog box. The procedure is somewhat different, but the result is the same. Select "Tools" from the pull down bar and then "Drawing Aids . . ." from the menu. This method will call the Drawing Aids dialog box shown in *Figure 2.1.* DDRMODES is the command that calls up this dialog box, so you can actually open it from the keyboard as well. Most dialog boxes are called by a command name beginning with DD, which stands for "Dynamic Dialog".

Look at the Drawing Aids dialog box. It contains some typical features, including check boxes, edit boxes, and radio buttons.

Figure 2.1.

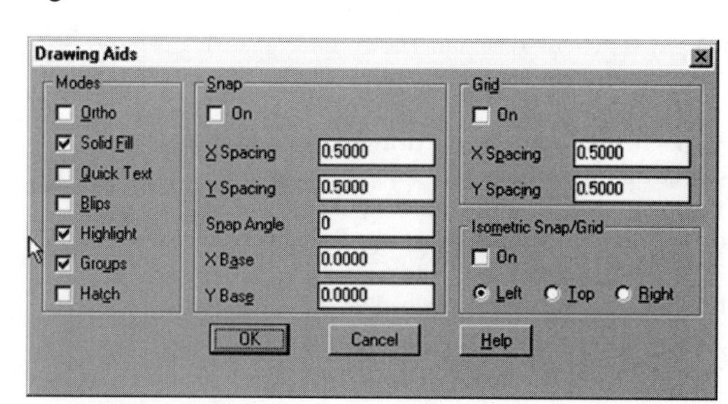

Check Boxes

You can turn ortho, snap, and grid on and off by moving the arrow inside the appropriate check box and pressing the pick button. A checked box is on, while an empty box is off. Notice that there are seven other check boxes in the area on the left under "Modes". We will only have use for the snap and grid settings at this point.

On the lower right you will see a box for isometric snap and grid control. We will have no use for these until Chapter 11. But notice the boxes in the lower line of the isometric snap/grid area labelled "Left", "Top", and "Right". These are examples of radio buttons, which are discussed in Task 3.

Also notice the "Help . . ." box at the bottom right. Again any box with an ellipsis (. . .) calls another dialog box that will overlay the current one. Picking this box would activate the HELP command, discussed later in this chapter.

Edit Boxes

The snap and grid settings are shown in edit boxes. Edit boxes contain text or numerical information that can be edited as you would in a text editor. You can highlight the entire box to replace the text, or point anywhere inside to do partial editing.

To change the grid or snap setting, use the following procedure:

1. Move the arrow into the box in the table where the change is to be made ("X Spacing" under "Snap" or "Grid").
2. Double click to highlight the entire number.
3. Type a new value and press enter. Notice that the Y spacing changes automatically when you enter the X spacing (see note following).
4. Click on the OK box at the bottom to confirm changes, or "Cancel" to cancel changes.

NOTE: The dialog box has places to set both X and Y spacing. It is unlikely that you will want to have a grid or snap matrix with different horizontal and vertical increments, but the capacity is there if you do. Also notice that you can change the snap angle. Setting the snap angle to 45, for example, would turn your snap and grid at a 45 degree angle. When you turn the grid or snapangle, you may also want to designate x and y base points around which to rotate. This is the function of the X and Y base point settings.

⌗ **Before going on to Task 2, use the dialog box to set the grid back to 0.5000.**

2.2 CHANGING THE SNAP

Whether you are typing or using one of your menus, the process for changing the snap setting is the same as changing the grid. In fact the two are similar enough to cause confusion. The grid is only a visual reference. It has no effect on selection of points. Snap is invisible, but it dramatically affects point selection. Grid and snap may or may not have the same setting.

⌗ **The grid should be set to 0.5000 from the previous task before you begin.**

⌗ **Type "snap" or open the Drawing Aids from the Tools menu.**

If you are typing, the prompt will appear like this, with options separated by slashes (/):

Snap spacing(X) or ON/OFF/Aspect/Rotate/Style <0.5000>:

⌗ **Type "1".**

This makes the snap setting larger than the grid setting. Move the cursor around the screen and you will see that you can only access half of the grid points. Clearly this type of arrangement is not too useful.

⌗ **Try other snap settings. Try 2, .25, and .125.**

⌗ **Set the snap to .05.**

Remember what happened when you tried to set the grid to .05?

⌗ **Move the cursor and watch the coordinate display.**

Observe how the snap setting is reflected in the available coordinates. How small a snap will AutoCAD accept?

⌗ **Try 0.005.**

Move the cursor and observe the coordinate display.

⌗ **Try 0.0005.**

You could even try 0.0001, but this would be like turning snap off, since you are only going to 4 decimal places anyway. The point is that unlike the grid which is limited by the size of your screen, you can set snap to any value you like.

⌗ **Finally, before you leave this task, set the snap to .25 and the grid to .5.**

This is the most efficient type of arrangement. With the grid set "coarser" than the snap you can still pick exact points easily, but the grid is not so dense as to be distracting.

NOTE: If you wish to keep snap and grid the same, set the grid to "0" in the Drawing Aids dialog box, or enter the GRID command and then type "s", for the Snap option. The grid will then change to match the snap and will continue to change any time you reset the snap. To free the grid, just give it its own value again using the GRID command or the dialog box.

2.3 CHANGING UNITS

⌗ **Type "Units"** (do not use the pull down menu yet).

Presto! The text window appears and covers the drawing window. We told you this would happen. The command area is too small to display the complete UNITS sequence, so the drawing window has been temporarily overlaid by the text window. What function key will bring it back?

What you now see looks like this:

Report Formats:	(Examples)
1. Scientific	1.55E + 01
2. Decimal	15.50
3. Engineering	1′-3.50″
4. Architectural	1′-3 1/2″
5. Fractional	15 1/2

With the exception of Engineering and Architectural formats, these formats can be used with any basic unit of measurement. For example, decimal mode works for metric units as well as English units.

Enter choice, 1 to 5 <2>:

⌖ **Type "2" or simply press enter, since the default system, and the one we will use is number 2, decimal units.**

Through most of this book we will stick to decimal units. Obviously, if you are designing a house you will want architectural units. If you are building a bridge you may want engineering-style units. You might want scientific units if you are doing research.

Whatever your application, once you know how to change units, you can do so easily and at any time. However, as a drawing practice you will want to choose appropriate units when you first begin work on a new drawing.

AutoCAD should now be showing the following prompt:

Number of digits to right of decimal point, (0 to 8) <4>:

We will use two-place decimals because they are practical and more common than any other choice.

⌖ **Type "2" in answer to the prompt for the number of decimal places you wish to use.**

AutoCAD now gives you the opportunity to change the units in which angles are measured. In this book we will use all of the default settings for angle measure, since they are by far the most common. If your application requires something different, the UNITS command or the DDUNITS dialog box is the place to change it.

The default system is standard degrees without decimals, measured counterclockwise, with 0 being straight out to the right (3 o'clock), 90, straight up (12 o'clock), 180 to the left (9 o'clock), and 270 straight down (6 o'clock).

⌖ **Press enter four times, or until the "Command:" prompt reappears.** Be careful not to press enter again, or the UNITS command sequence will be repeated.

Looking at the coordinate display, you should now see values with only two digits to the right of the decimal. This setting will be standard in this book.

We suggest that you experiment with other choices in order to get a feel for the options that are available to you. You should also try the Units Control dialog box described next.

The Units Control Dialog Box

Activate the Units Control dialog box by picking "Units . . ." from the Format pull down menu. You will see the command DDUNITS entered on the command line, and the dialog box shown in *Figure 2.2* will appear. This box shows all of the settings that are also available through the UNITS command sequence. There are two dialog box features here that we have not discussed previously.

Radio Buttons

First are the radio buttons in the columns labelled "Units" and "Angles". Radio buttons are used with lists of settings that are mutually exclusive. You should see that the Decimal button is shaded in the format column. All other buttons in this column are not shaded. You can switch settings by simply picking another button, but you can have only one button on at a time. Radio buttons are used here because you can use only one format at a time.

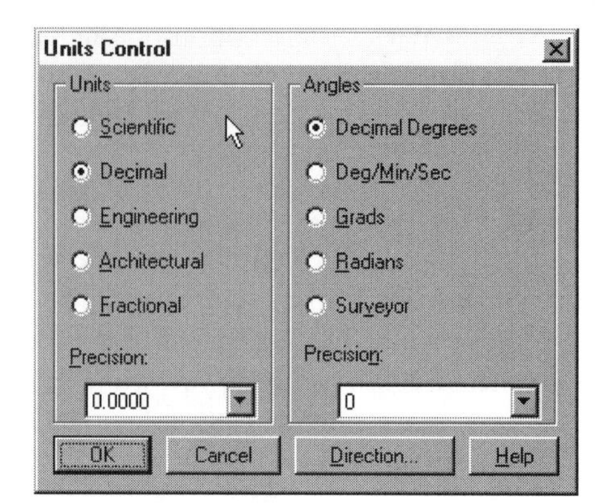

Figure 2.2.

Pop Down Lists

The second new feature is the pop down lists at the bottom of the units and angles columns. You should try these out to see how they function. You can change the precision (number of place values shown) by picking the arrow at the right of the box and then picking a setting, such as "0.000" for three place decimals, from the list that appears.

Experiment as much as you like with the dialog box. None of your changes will be reflected in the drawing until you click on the OK button or press enter on the keyboard.

When you are through experimenting, be sure to leave your units set for two-place decimals, your angle measure set for zero-place decimal degrees, and your angle zero direction set to East.

NOTE: All dialog boxes can be moved on the screen. This is done by clicking in the gray title area at the top of the dialog box, holding down the pick button, and dragging the box across the screen.

2.4 DRAWING CIRCLES GIVING CENTER POINT AND RADIUS

Circles can be drawn by giving AutoCAD three points on the circle's circumference, two points that determine a diameter, two tangents and a radius, a center point and a radius, or a center point and a diameter.
In this chapter we will use the latter two options.

We will begin by drawing a circle with radius 3 and center at the point (6,5). Then we will draw two smaller circles centered at the same point. Later we will erase them using the ERASE command.

Grid should be set to .5, snap to .25, and units to two-place decimal.

Type "c" , select Circle from the Draw menu, or select the Circle tool from the Draw toolbar, illustrated in *Figure 2.3*.

The prompt that follows will look like this:

3P/2P/TTR/<Center point>:

Type coordinates or point to the center point of the circle you want to draw. In our case it will be the point (6,5).

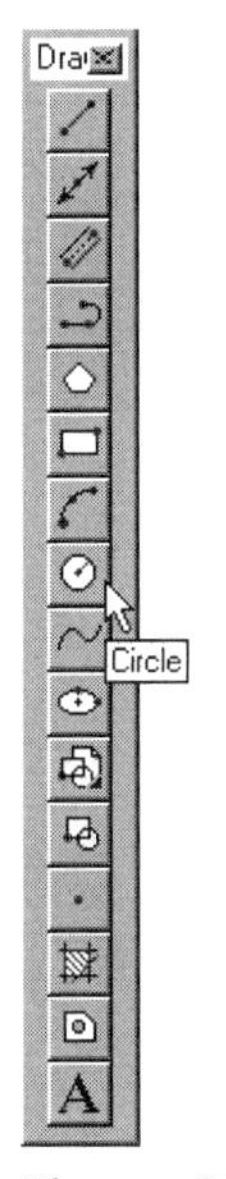

Figure 2.3.

AutoCAD assumes that a radius or diameter will follow and shows the following prompt:

Diameter/<Radius>:

If we type or point to a value now, AutoCAD will take it as a radius, since that is the default.

Type "3" or show by pointing that the circle has a radius of 3.

Notice how the rubber band works to drag out your circle as you move the cursor.

If your coordinate display is not showing polar coordinates, press F6 once or twice until you see something like "3.00 < 0, 0.00". When you are ready press the pick button to show the radius end point.

You should now have your first circle complete. Next,

Draw two more circles using the same center point, radius method. They will be centered at (6,5) and have radii of 2.50 and 2.00.

The results are illustrated in *Figure 2.4*.

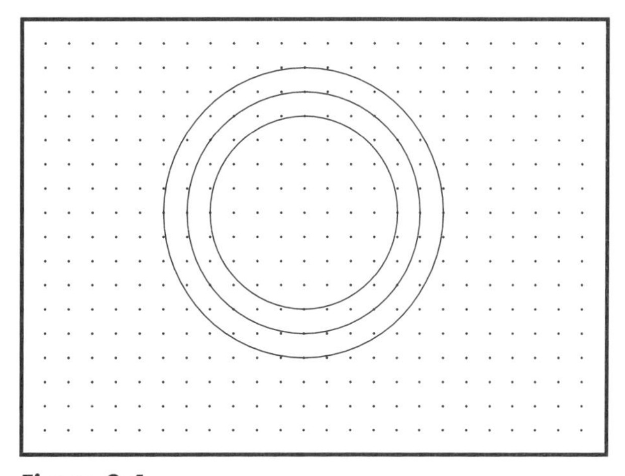

Figure 2.4.

2.5 DRAWING CIRCLES GIVING CENTER POINT AND DIAMETER

We will draw three more circles centered on (6,5) having diameters of 2, 1.5, and 1. Drawing circles this way is almost the same as the radius method, except you will not use the default, and you will see that the rubber band works differently.

Press enter or the space bar to repeat the CIRCLE command.
Indicate the center point (6,5) by typing coordinates or pointing.
Answer the prompt with a "d", for diameter.

Notice that the cross hairs are now outside the circle you are dragging on the screen (see *Figure 2.5*). This is because AutoCAD is looking for a diameter, but the last point you gave was a center point. So the diameter is being measured from the

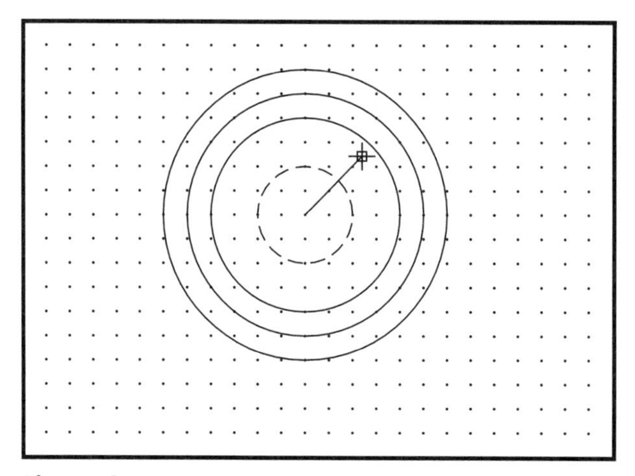

Figure 2.5.

center point out, twice the radius. Move the cursor around, in and out from the center point, to get a feel for this.

⌖ **Point to a diameter of 2.00, or type "2".**

You should now have four circles.

⌖ **Draw two more circles with diameters of 1.0 and 1.5**

When you are done, your screen should look like *Figure 2.6.*

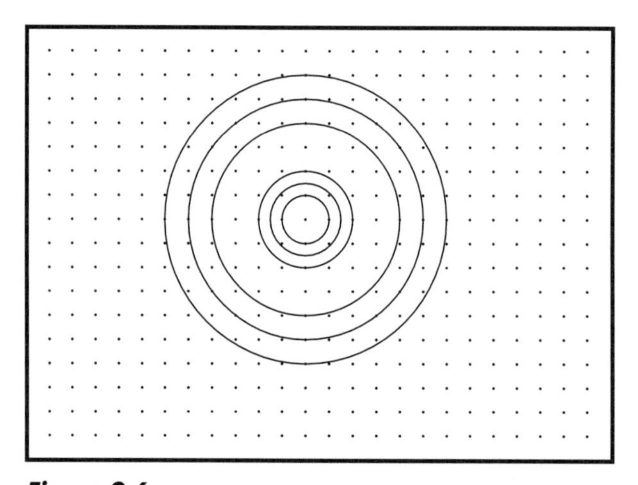

Figure 2.6.

2.6 ACCESSING AUTOCAD HELP FEATURES

The Release 14 HELP command gives access to a complete library of AutoCAD references and information. The procedures are standard Windows 95 help procedures and give you access to the *AutoCAD Command Reference,* the *AutoCAD User's Guide,* the *AutoCAD Customization Guide,* the *AutoCAD Installation Guide,* and the *Release 14 Tutorial.* In this task we will focus on the use of the Index feature, which pulls information from all of the references, depending on the topic you select. For demonstration, we will look at the CIRCLE command.

To begin you should be at the command prompt.

HELP is context sensitive, meaning that it will go directly to the *AutoCAD Command Reference* if you ask for help while in the middle of a command sequence. You should try this later.

Press F1, open the Help menu and select AutoCAD Help topics . . ., or select the question mark at the right end of the Standard toolbar.

This will open the AutoCAD Help dialog box shown in *Figure 2.7.* If the Contents tab is showing, you will see the list of available references. These contain the same information as the actual AutoCAD manuals and you can look through their contents as you would with the books. It is usually quicker to use the index, however.

If necessary, click on the Index tab.

You should see the Index as shown in *Figure 2.7.* The list of topics is very large so it is rare that you will use the scroll bar on the right. Most often you will type in a command or topic. The list will update as you type, so often you will not need to type the complete word or command.

Type "Cir".

If you do this slowly, one letter at a time, you will see how the index follows along. When you have typed "cir" you will be at the Circle (Draw menu), options entry, adding the rest of the letters "cle" will have no further effect.

The entry we want is the CIRCLE command entry which is just below the Draw menu options.

Double click on "CIRCLE command" or highlight it and then press enter or click on display.

This will call up a second smaller dialog box with three options. We want the first option.

Double click on "CIRCLE command" again, or press enter or click on display.

Figure 2.7.

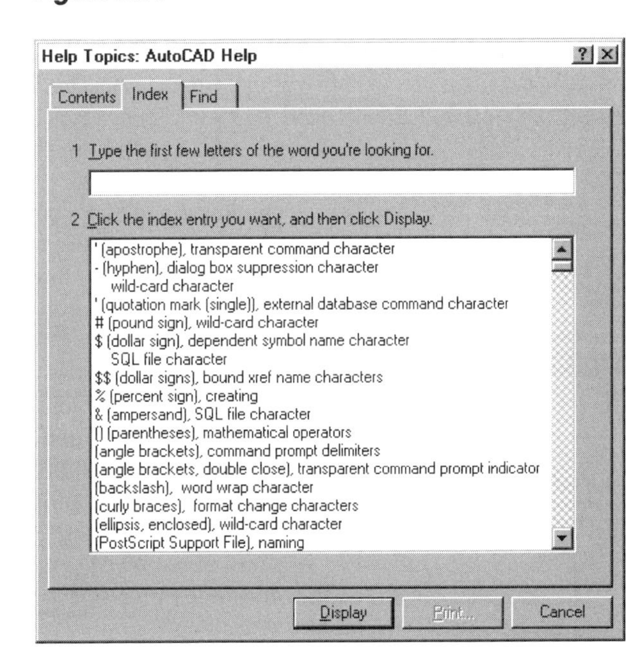

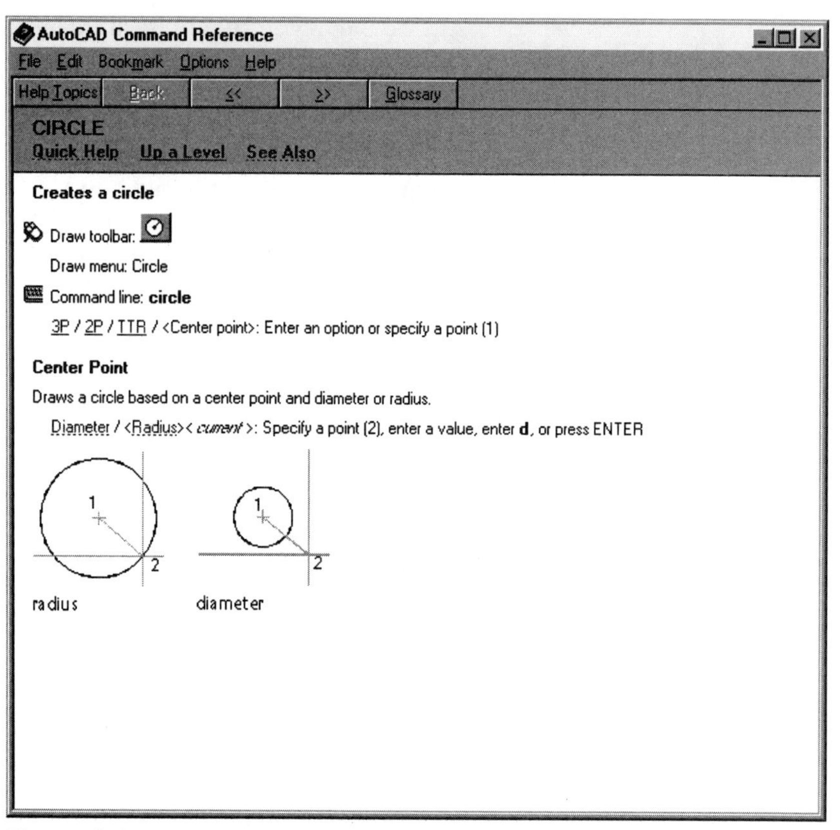

Figure 2.8.

AutoCAD will display the CIRCLE command page from the *AutoCAD Command Reference* as shown in *Figure 2.8.* Additional information on items on this page is available for words shown in green.

⌖ Click on "TTR".

This will call up the page for the Tangent, Tangent, Radius option of the CIRCLE command, shown in *Figure 2.9.*

⌖ To exit HELP, click on the close button, the x in the upper right corner of the help window.

This terminates the HELP command and brings you back to the command prompt. The HELP feature contains a vast amount of information. You are encouraged to access it at any time to expand your knowledge of AutoCAD.

2.7 USING THE ERASE COMMAND

AutoCAD allows for many different methods of editing and even allows you to alter some of the basics of how edit commands work. Fundamentally, there are two different sequences for using most edit commands. These are called the Noun/Verb and the Verb/Noun methods.

In earlier versions of AutoCAD, most editing was carried out in a verb/noun sequence. That is, you would enter a command, such as ERASE (the verb), then select

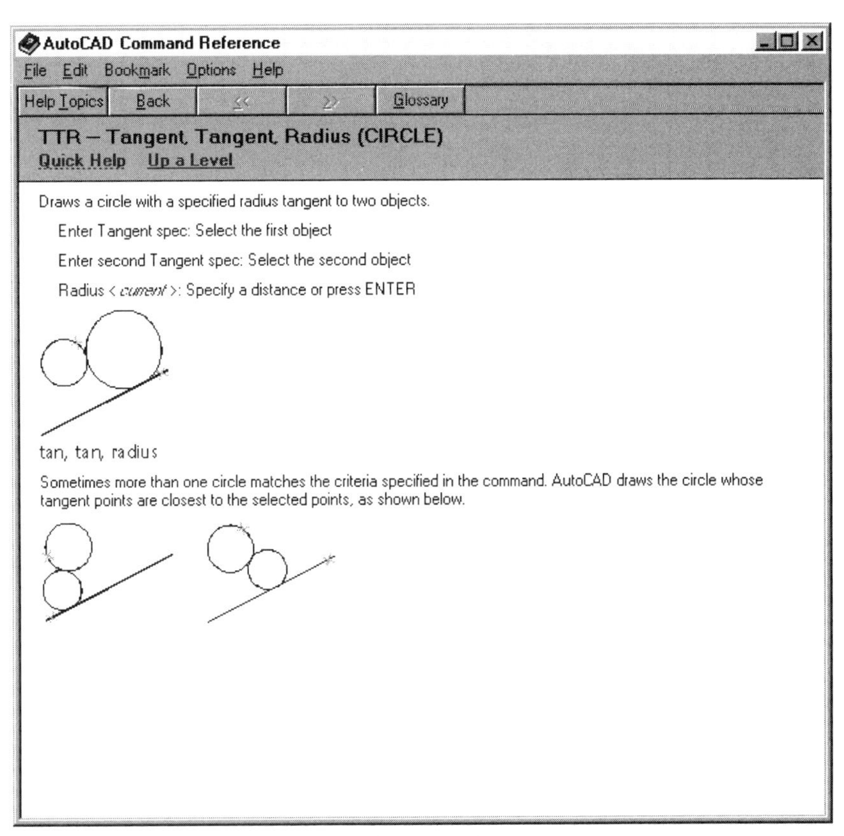

Figure 2.9.

objects (the nouns), and finally press enter to carry out the command. This method is still perfectly reasonable and effective, but AutoCAD now allows you to reverse the verb/noun sequence. You can use either method as long as "Noun/Verb" selection is enabled in your drawing.

In this task we will explore the traditional verb/noun sequence and then introduce the noun/verb or "pick first" method along with some of the many methods for selecting objects.

Verb/Noun Editing

⌗ To begin this task you should have the six circles on your screen, as shown previously in *Figure 2.6.*

We will use verb/noun editing to erase the two outer circles.

⌗ Type "e", select Erase from the Modify menu, or select the Erase tool from the Modify toolbar, as shown in *Figure 2.10.*

The cross hairs will disappear, but the pickbox will still be on the screen and will move when you move your cursor.

Figure 2.10.

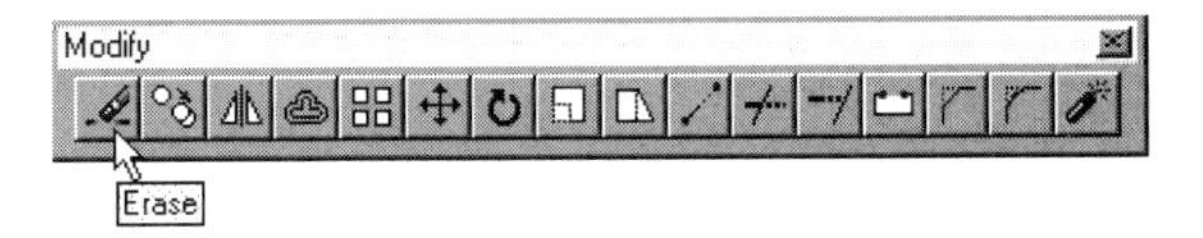

Also notice the command area. It should be showing this prompt:

Select objects:

This is a very common prompt. You will find it in all edit commands and many other commands as well.

Move your cursor so that the outer circle crosses the pickbox.

TIP: In many situations you may find it convenient or necessary to turn snap off (F9) while selecting objects, since this gives you more freedom of motion.

Press the pick button.

The circle will be highlighted (dotted). This is how AutoCAD indicates that an object has been selected for editing. It is not yet erased, however. You can go on and add more objects to the selection set and they, too, will become dotted.

Use the box to pick the second circle. It too should now be dotted.

Press enter, the space bar, or the enter equivalent button on your pointing device to carry out the command.

This is typical of the verb/noun sequence in most edit commands. Once a command has been entered and a selection set defined, a press of the enter key is required to complete the command. At this point the two outer circles should be erased.

Noun/Verb Editing

Now let's try the noun/verb sequence.

Type "u" to undo the ERASE and bring back the circles.

Use the pickbox to select the outer circle.

The circle will be highlighted, and your screen should now resemble *Figure 2.11.* Those little blue boxes are called "grips". They are part of AutoCAD's auto-editing system, which we will begin exploring in Chapter 3. For now you can ignore them.

Pick the second circle in the same fashion.

The second circle will also become dotted and more grips will appear.

Type "e" , select Erase from the Modify menu, or the Erase tool from the Modify toolbar.

Your two outer circles will disappear as soon as you press enter or pick the tool.

Figure 2.11.

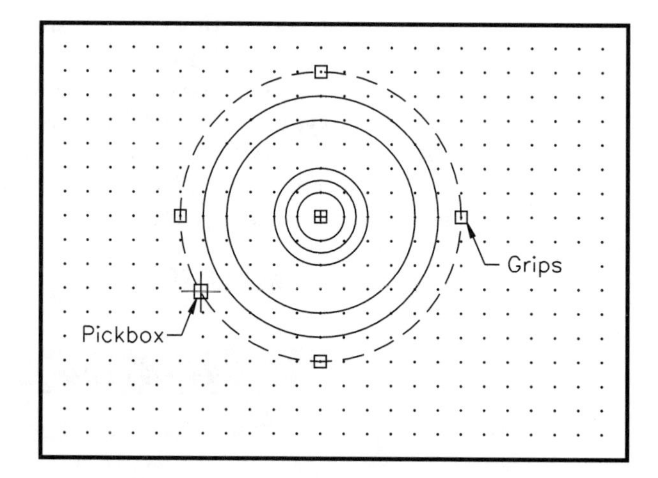

The two outer circles should now be gone. As you can see there is not a lot of difference between the two sequences. One difference that is not immediately apparent is that there are numerous selection methods available in the older verb/noun system that cannot be activated when you pick objects first. We will get to other object select methods momentarily, but first try out the OOPS command.

OOPS!

⌖ **Type "oops" and watch the screen.**

If you have made a mistake in your erasure, you can get your selection set back by typing "oops." OOPS is to ERASE as REDO is to UNDO. You can use OOPS to undo an ERASE command, as long as you have not done another ERASE in the meantime. In other words, AutoCAD only saves your most recent ERASE selection set.

You can also use U to undo an ERASE, but notice the difference: U simply undoes the last command, whatever it might be; OOPS works specifically with ERASE to recall the last set of erased objects. If you have drawn other objects in the meantime, you can still use OOPS to recall a previously erased set. But if you tried to use U, you would have to backtrack, undoing any newly drawn objects along the way.

Other Object Selection Methods

You can select individual entities on the screen by pointing to them one by one, as we have done above, but in complex drawings this will often be inefficient. AutoCAD offers a variety of other methods, all of which have application in specific drawing circumstances. In this exercise we will select circles by the "windowing" and "crossing" methods, by indicating "last" or "L", meaning the last entity drawn, and by indicating "previous" or "P" for the previously defined set. There are also options to add or remove objects from the selection set and other variations on windowing and crossing. We suggest that you study *Figure 2.12* to learn about other methods. The number of selection options available may seem a bit overwhelming at first, but time learning them will be well spent. These same options appear in many AutoCAD editing commands (MOVE, COPY, ARRAY, ROTATE, MIRROR) and should become part of your CAD vocabulary.

Selection by Window

Window and crossing selections, like individual object selection, can be initiated without entering a command. In other words, they are available for noun/verb selection. Whether you select objects first or enter a command first, you can force a window or crossing selection simply by picking points on the screen that are not on objects. AutoCAD will assume you want to select by windowing or crossing and will ask for a second point.

Let's try it. We will show AutoCAD that we want to erase all of the inner circles by throwing a temporary selection window around them. The window will be defined by two points moving left to right that serve as opposite corners of a rectangle. Only entities that lie completely within the window will be selected. See *Figure 2.13.*

⌖ **Pick point 1 at the lower left of the screen, as shown. Any point in the neighborhood of (3.5,1) will do.**

AutoCAD will prompt for another corner:

Other corner:

⌖ **Pick point 2 at the upper right of the screen, as shown.** Any point in the neighborhood of (9.5,8.5) will do. To see the effect of the window, be sure that it crosses the outside circle as in *Figure 2.13.*

OBJECT SELECTION METHOD	DESCRIPTION	ITEMS SELECTED
(W) WINDOW		THE ENTITIES WITHIN THE BOX
(C) CROSSING		THE ENTITIES CROSSED BY OR WITHIN THE BOX
(P) PREVIOUS		THE ENTITIES THAT WERE PREVIOUSLY PICKED
(L) LAST		THE ENTITY THAT WAS DRAWN LAST
(R) REMOVE		REMOVES ENTITIES FROM THE ITEMS SELECTED SO THEY WILL NOT BE PART OF THE SELECTED GROUP
(A) ADD		ADDS ENTITIES THAT WERE REMOVED AND ALLOWS FOR MORE SELECTION AFTER THE USE OF REMOVE
ALL		ALL ENTITIES CURRENTLY VISIBLE ON THE DRAWING
(F) FENCE		THE ENTITIES CROSSED BY THE FENCE
(WP) WPOLYGON		THE ENTITIES WITHIN THE THE POLYGON
(CP) CPOLYGON		THE ENTITIES CROSSED BY OR WITHIN THE PLOYGON

Figure 2.12.

Figure 2.13.

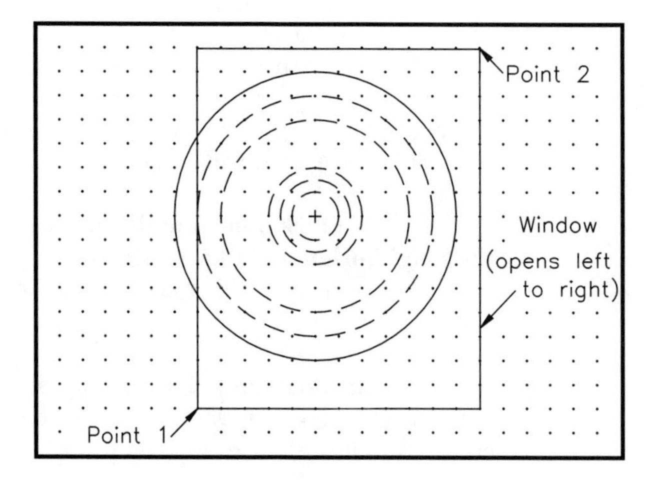

⌗ Type "e" or select the Erase tool.

The inner circles should now be erased.

⌗ Type or select "oops" to retrieve the circles once more. Since ERASE was the last command, typing "U" will work equally well.

Selection by Crossing Window

Crossing is an alternative to windowing that is useful in many cases where a standard window selection could not be performed. The selection procedure is the same, but a crossing box opens to the left instead of to the right and all objects that cross the box will be chosen, not just those that lie completely inside the box.

We will use a crossing box to select the inside circles.

⌗ Pick point 1 close to (8.0,3.0) as in *Figure 2.14.*

AutoCAD prompts:

Other corner:

⌗ Pick a point near (4.0,7.0).

This point selection must be done carefully in order to demonstrate a crossing selection. Notice that the crossing box is shown with dotted lines, whereas the window box was shown with solid lines.

Also, notice how the circles are selected: those that cross and those that are completely contained within the box, but not those that lie outside.

At this point we could enter the ERASE command to erase the circles, but instead we will demonstrate how to use the Esc key to cancel a selection set.

⌗ Press the Esc key on your keyboard.

This will cancel the selection set. The circles will no longer be highlighted, but you will see that the grips are still visible. To get rid of the grips you will need to cancel again.

⌗ Press Esc again.

The grips should now be gone as well.

Selecting the "Last" Entity

AutoCAD remembers the order in which new objects have been drawn during the course of a single drawing session. As long as you do not leave the drawing, you can

Figure 2.14.

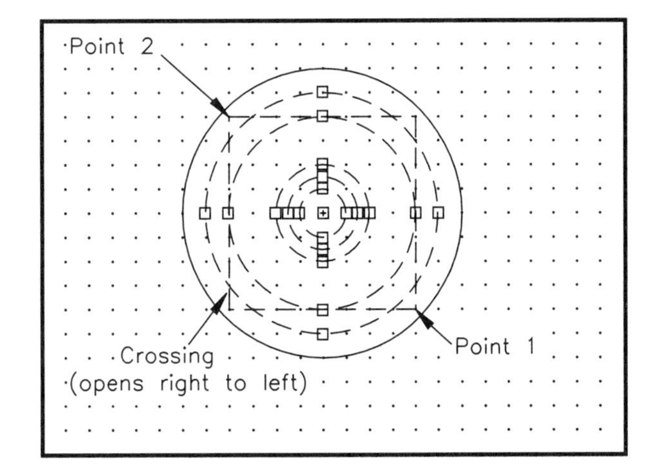

select the last drawn entity using the "last" option. If you leave the drawing and return later, this information will no longer be available.

☩ **Type "e" or select the Erase tool.**

Notice that there is no way to specify "last" before you enter a command. This option is only available as part of a command procedure. In other words, it only works in a verb/noun sequence.

☩ **Type "L".**

The inner circle should be highlighted.

☩ **Press enter to carry out the command.**

The inner circle should be erased.

There are several more object selection methods, all of which may be useful to you in a given situation. Some of the most useful options are described and illustrated in *Figure 2.12,* which you should study before moving on.

2.8 PLOTTING OR PRINTING A DRAWING

AutoCAD's printing and plotting capabilities are extensive and complex. In this book we will introduce you to them a little bit at a time. These presentations are intended to get your work out on paper efficiently. You will find discussions of plotting and printing in Chapters 2, 3, 4, 5, 6, and 8, before the review material and drawings.

In this chapter, we will show you how to do a very simple plot procedure. This procedure assumes that you do not have to change devices or configuration details. It should work adequately for the drawings in this chapter. One of the difficulties is that different types of plotters and printers work somewhat differently. We will try to present procedures that will work on whatever equipment you are using, assuming that it is appropriately configured to begin with.

We suggest that you now open the Plot Configuration dialog box and work through this section without actually plotting anything. Then come back to the general procedure after you have completed one of the drawings in this chapter that you want to print out.

☩ **Type Ctrl-p, select the Print tool from the Standard toolbar, or select "Print . . ." from the File menu.**

Any of these methods will call up the "Plot Configuration" dialog box illustrated in *Figure 2.15.* You will become very familiar with it as you work through this book. It is one of the most important working spaces in AutoCAD. It contains many options and will call many sub dialogs, but for now we are going to look at only two settings.

Look at the box on the lower left labeled "Additional Parameters". The line of radio buttons on the left allows you to tell AutoCAD what part of your drawing you want to plot. For our purposes we will use "Window". For this exercise you will define a window around an object you have drawn.

Windowing allows you to plot any portion of a drawing simply by defining the window. AutoCAD will base the size and placement of the printed drawing on the window you have defined on the screen. For now, any object will do.

☩ **Click on "Window..." at the bottom middle of the Additional Parameters box.**

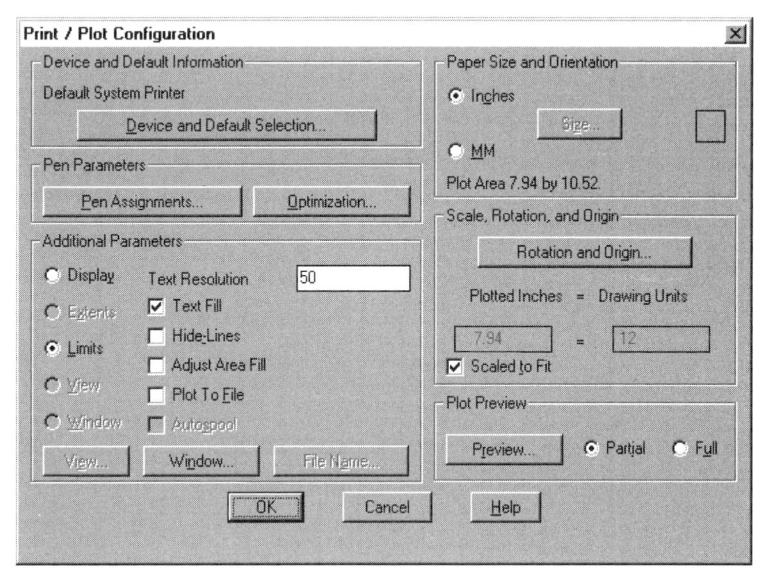

Figure 2.15.

This will call up the Window Selection dialog box illustrated in *Figure 2.16.* It contains the x and y coordinates of the two corners of your window selection. To ensure that you get the window you want, you need to pick the two points, or type in the coordinates if you know them. We will pick points, using *Drawing 2.1* to illustrate.

⌗ Click on "Pick <" at the top left of the dialog box.

AutoCAD will temporarily close the Plot Configuration dialog box so that you can pick points. You will be prompted:

First corner:

⌗ Pick a point similar to point 1 in *Figure 2.17.* For this drawing, a point in the neighborhood of (1.5,.25) will do.

⌗ Pick a point similar to point 2 in *Figure 2.17.* For this drawing, a point in the neighborhood of (10.5,8.75) will do.

As soon as you have picked the second point, AutoCAD will display the Window Selection box again, showing the coordinates you have chosen.

Figure 2.16.

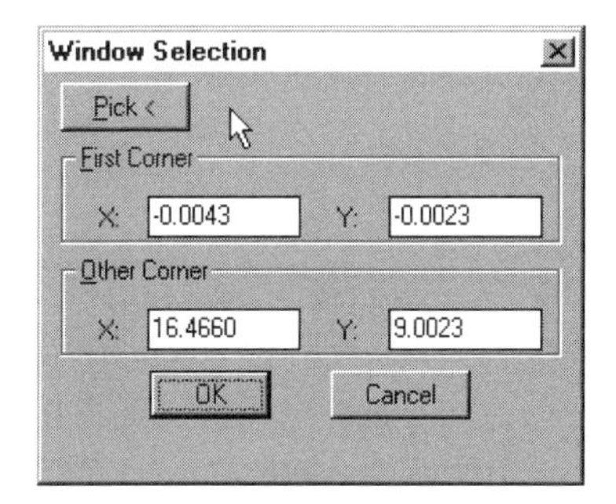

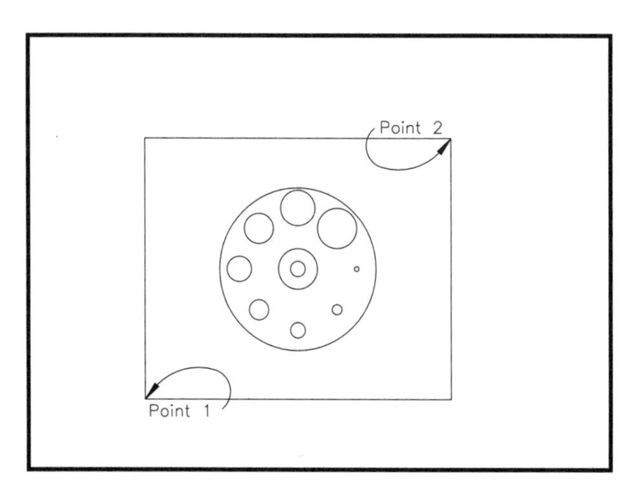

Figure 2.17.

⌖ **Click on "OK".**

This will bring you back to the Plot Configuration dialog box. You are essentially done at this point. But before plotting, there is one thing to check. Look at the box at the middle right labeled "Scale, Rotation, and Origin". It should not be necessary to worry about any of these parameters at this point. If the "Scaled to Fit" box is checked, as it should be by default, then AutoCAD will plot your drawing at maximum size based on paper size and the window you have defined.

⌖ **If "Scaled to Fit" is not checked, click in the box to check it.**

⌖ **If you are not actually going to print a drawing at this point, click on Cancel to close the Plot Configuration dialog box.**

At this point the exercise is complete. You should come back to it when you have a drawing ready to plot.

⌖ **If you are ready to proceed with a plot, look at your printer or plotter. Make sure that it is on, ready, and that the paper is ready to go.**

⌖ **Click on "OK".**

This sends the drawing information out to be printed. You can sit back and watch the plotting device at work. If you need to cancel for any reason, click the Cancel button.

For now, that's all there is to it. If your plot does not look perfect, if it is not centered, for example, don't worry, you will learn all you need to know about plotting in later chapters.

2.9 REVIEW MATERIAL

Questions

1. Which is likely to have the smaller setting GRID or SNAP? Why? What happens if the settings are reversed?
2. Why do some dialog box areas use check boxes while others use radio buttons?
3. How do you switch from decimal units to architectural units?
4. Where is 0 degrees located in AutoCAD's default units setup? Where is 270 degrees? Where is –45 degrees?

5. How do you enter the CIRCLE command?
6. Why does the rubber band touch the perimeter of the circle when you are using the radius option, but not when you are using the diameter option?
7. How does AutoCAD know when you want a crossing selection instead of a window selection?
8. What is the difference between a "Last" selection and a "Previous" selection?
9. What is the difference between noun/verb and verb/noun editing?
10. How do you access the AutoCAD HELP index.

COMMANDS

Tools	Draw	Modify
DDRMODES	CIRCLE	ERASE
GRID		
SNAP	Format	Help
	UNITS	HELP

Drawing Problems

1. Use HELP to look up the CMDDIA system variable. If this variable were changed to 0, instead of 1, what would happen when you select the Print tool?
2. Use the 3P option to draw a circle that passes through the points (3,3), (6,6), and (9,3).
3. Using the 2P option, draw a second circle with a diameter from (9,3) to (12,3).
4. Using the TTR option, draw a third circle tangent to the first two with a radius of 1.50.

PROFESSIONAL SUCCESS

Ensuring Your File Is Compatible with All AutoCAD Releases

Imagine this scenario: You work all night at home on an important AutoCAD assignment, which must be plotted and turned in the following day. In the computer lab, you access AutoCAD, and try to load your file from the floppy disk you used at home. You panic when you're unable to load your file, and wonder what has gone wrong . . .

Chances are that you are using two different releases of AutoCAD: one at home, and another, older one in the computer lab. While all versions of AutoCAD are downward compatible (files created on older versions can be opened and interpreted by newer releases), files created using the most recent releases of AutoCAD may not open on older versions. Luckily,there is a solution: Save your file using the "Save As" command, and choose the appropriate older format in the pop-up dialog box. You should never face the scenario described if you choose a format compatible with the oldest release of AutoCAD that you are using.

This issue is not unique to AutoCAD. Many other popular software packages, including Microsoft Office, allow you to save files in various formats, so that they can be read and modified by older releases of the software. It is good practice to always keep track of which software releases you are using, and to make sure that you save your files in a format compatible with all of them. This precaution will increase the chances that your files will open whenever you try to access them.

DRAWING 2—1

Aperture Wheel

This drawing will give you practice drawing circles using the center point, radius method. Refer to the table below the drawing for radius sizes. With snap set at .25, some of the circles can be drawn by dragging and pointing. Other circles have radii that are not on a snap point. These circles can be drawn easily by typing in the radius.

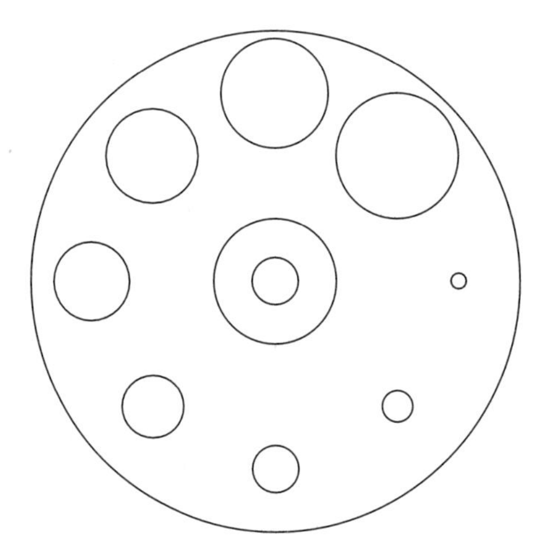

Drawing Suggestions

GRID = .50

SNAP = .25

- A good sequence for doing this drawing would be to draw the outer circle first, followed by the two inner circles (h and c). These are all centered on the point (6.00,4.50). Then begin at circle a and work around clockwise, being sure to center each circle correctly.
- Notice that there are two circles c and two h. This simply indicates that the two circles having the same letter are the same size.
- Remember, you may type any value you like and AutoCAD will give you a precise graphic image, but you cannot always show the exact point you want with a pointing device. Often it is more efficient to type a few values than to turn snap off or change its setting for a small number of objects.

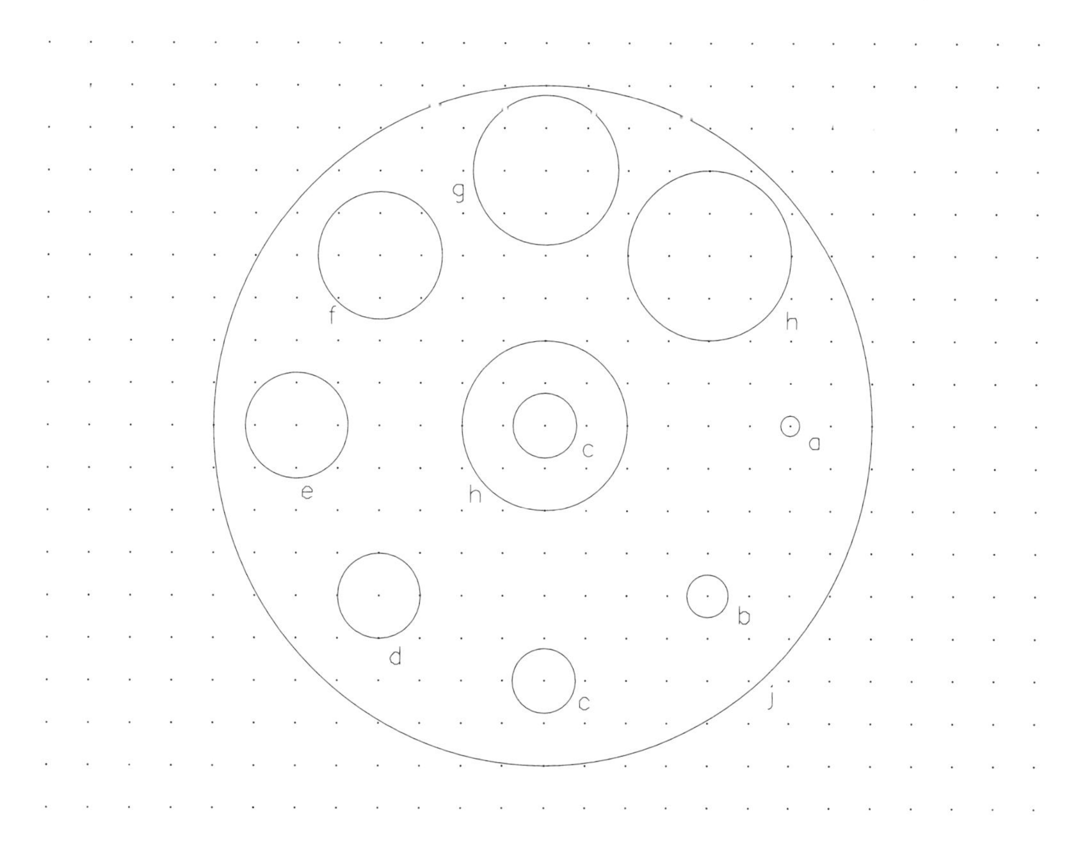

LETTER	a	b	c	d	e	f	g	h	j
RADIUS	.12	.25	.38	.50	.62	.75	.88	1.00	4.00

APERTURE WHEEL

Drawing 2-1

DRAWING 2—2

Roller

This drawing will give you a chance to combine lines and circles and to use the center point, diameter method. It will also give you some experience with smaller objects, a denser grid, and a tighter snap spacing.

NOTE: Even though units are set to show only two decimal places, it is important to set the snap using three places (.125) so that the grid is on a multiple of the snap (.25 = 2 × .125). AutoCAD will show you rounded coordinate values, like .13, but will keep the graphics on target. Try setting snap to either .13 or .12 instead of .125, and you will see the problem for yourself.

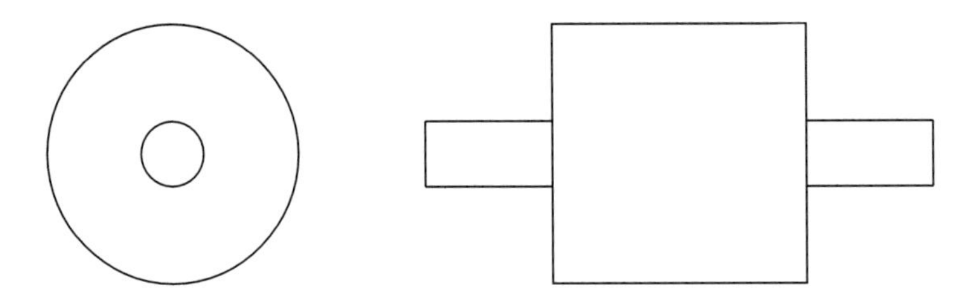

Drawing Suggestions

GRID = .25
SNAP = .125

- The two views of the roller will appear fairly small on your screen, making the snap setting essential. Watch the coordinate display as you work and get used to the smaller range of motion.
- Choosing an efficient sequence will make this drawing much easier to do. Since the two views must line up properly, we suggest that you draw the front view first, with circles of diameter .25 and 1.00, and then use these circles to position the lines in the right side view.
- The circles in the front view should be centered in the neighborhood of (2.00,6.00). This will put the upper left-hand corner of the 1 × 1 square at around (5.50,6.50).

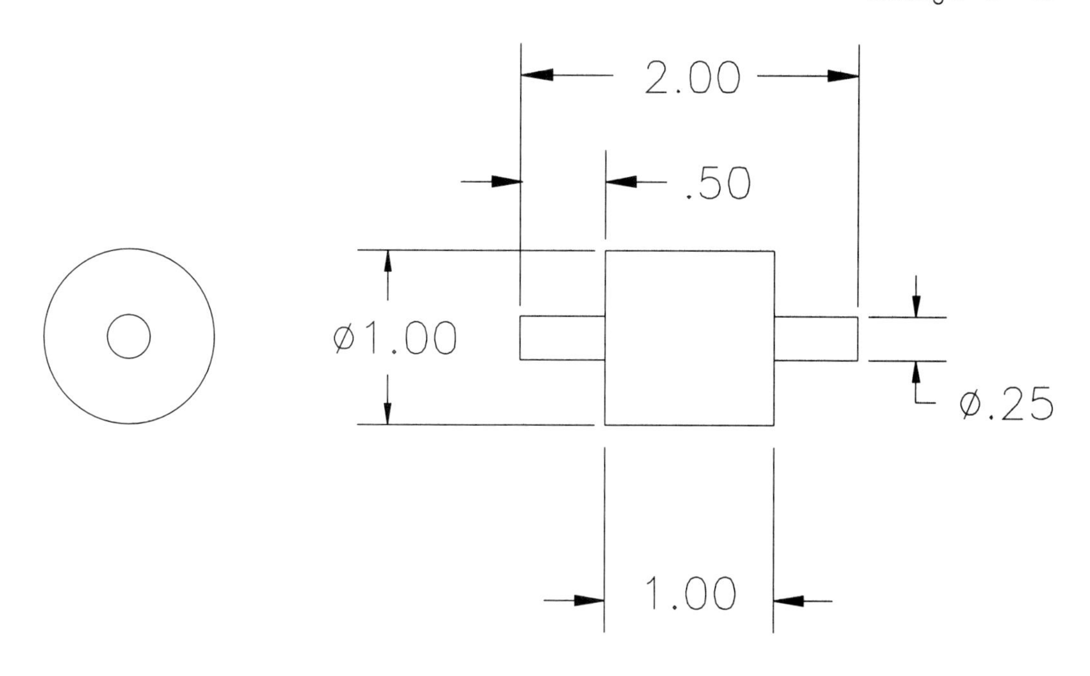

ROLLER
Drawing 2-2

DRAWING 2—3

Switchplate

This drawing is similar to the last one, but the dimensions are more difficult, and a number of important points do not fall on the grid. It will give you practice using grid and snap points and the coordinate display. Refer to the table below the drawing for dimensions of the circles, squares, and rectangles inside the 7 × 10 outer rectangle. The placement of these smaller figures is shown by the dimensions on the drawing itself.

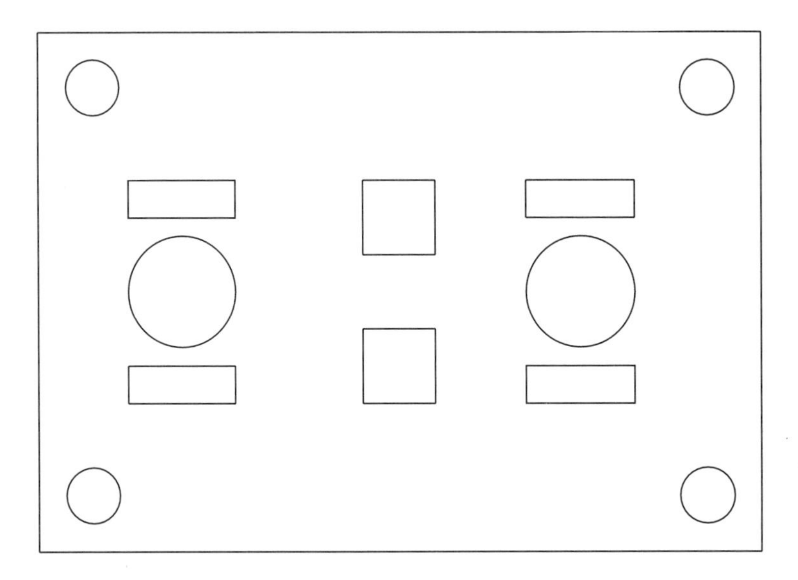

Drawing Suggestions

GRID = .50
SNAP = .25

- Turn ortho on to do this drawing.
- A starting point in the neighborhood of (1,1) will keep you well positioned on the screen.

Guide Points with Dist The squares, rectangles, and circles in this drawing can be located easily using DIST to set up guide points. For example, set a first point at the lower left corner of the outer rectangle as in Reference 2.4b. Then set the second point at 1.25 to the right along the bottom of the rectangle. Now repeat the DIST command and use this second point as the new first point. Set the new second point 2.00 up, and you will have a blip right where you want to begin the c rectangle. Look for other places to use this technique in this drawing.

The Rectang Command The RECTANG command is a quick and easy way to draw rectangles. Type or select the command and then pick two points at opposite corners of the rectangle you want to draw. It is just like creating a selection window, but the window remains as a single entity called a polyline. Polylines are discussed in Chapter 10. Type "Rectang" or select the Rectangle tool from the Draw toolbar.

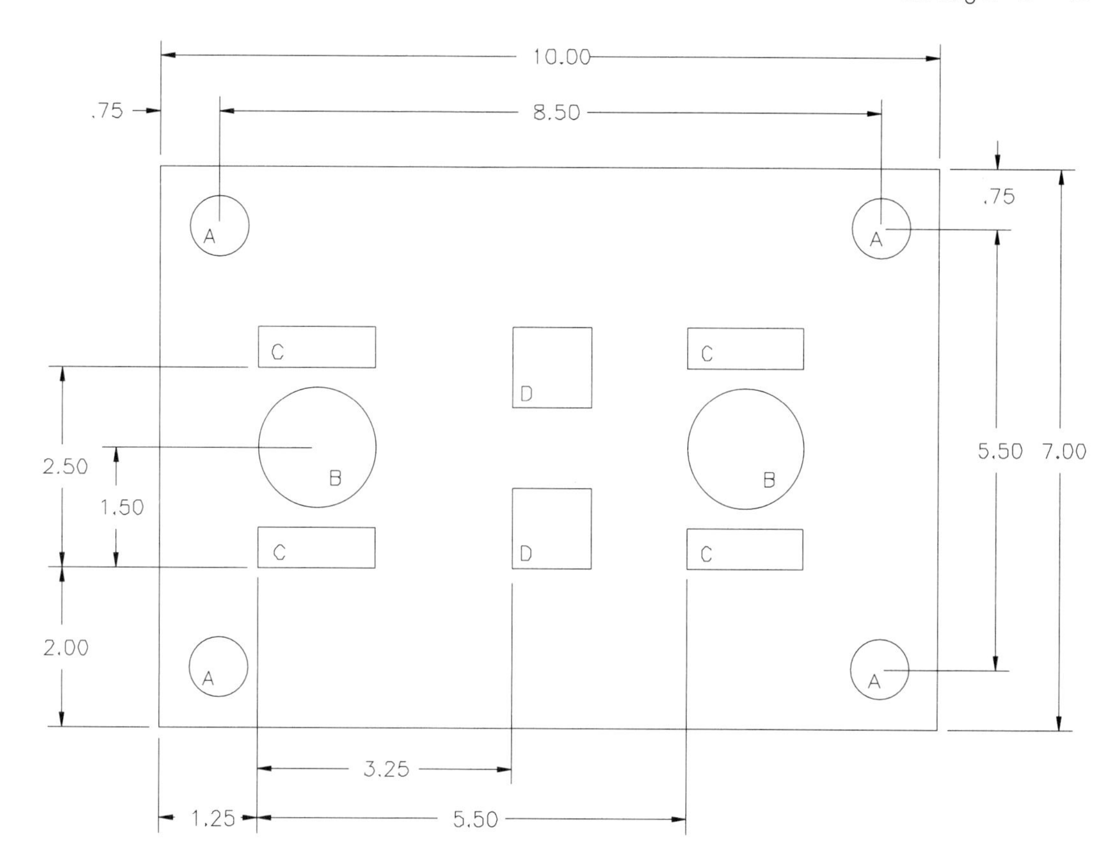

HOLE	SIZE
A	⌀.75
B	⌀1.50
C	.50 H x 1.50 W
D	1.00 SQ

SWITCH PLATE
Drawing 2-3

3

Layers, Colors, and Linetypes

OVERVIEW

So far all the drawings you have done have been on a single white layer called "0". In this chapter you will create and use three new layers, each with its own associated color and linetype.

You will also learn to FILLET and CHAMFER the corners of previously drawn objects, to magnify portions of a drawing using the ZOOM command, and to move between adjacent portions of a drawing with the PAN command.

SECTIONS

OBJECTIVES

After reading this chapter, you should be able to:

- Create three new layers.
- Assign colors to layers.
- Assign linetypes to layers.
- Change the current layer.
- FILLET the corners of a square.
- CHAMFER the corners of a square.
- ZOOM in and out using Window, Previous, and All.
- Use realtime PAN and ZOOM.
- Use the Plot Preview feature.
- Draw a Mounting Plate.
- Draw a Bushing.
- Draw a Half Block.

3.1 CREATING NEW LAYERS

Layers allow you to treat specialized groups of entities in your drawing separately from other groups. For example, all of the dimensions in this book were drawn on a special dimension layer so that we could turn them on and off at will. We turned off the dimension layer in order to prepare the reference drawings for chapters 1 through 7, which are shown without dimensions. When a layer is turned off, all the objects on that layer become invisible, though they are still part of the drawing database and can be recalled at any time.

It is common to put dimensions on a separate layer, and there are many other uses of layers as well. Fundamentally, layers are used to separate colors and linetypes, and these in turn take on special significance depending on the drawing application. It is standard drafting practice, for example, to use small, evenly spaced dashes to represent objects or edges that would in reality be hidden from view.

On a CAD system with a color monitor, these hidden lines can also be given their own color to make it easy for the designer to remember what layer he or she is working on.

In this book we will use a simple and practical layering system, most of which will be presented in this chapter. You should remember that there are many systems in use, and countless possibilities. AutoCAD allows as many as 256 different colors and as many layers as you like.

You should also be aware that linetypes and colors are not restricted to being associated with layers. It is possible to mix linetypes and colors on a single layer. But while this may be useful for certain applications, we do not recommend it at this point.

⌗ Open a new drawing using the Start from Scratch option to ensure that you are using the same defaults as those used in this chapter.

The Layer Control Dialog Box (DDLMODES)

In this instance we will introduce the dialog box procedures first because they are more efficient than the command line methods. The main advantage of using the dialog box is that a table of layers will be displayed in front of you as you make changes, and you will be able to make several changes at once.

⌗ Select the Layer tool from the Object Properties toolbar, as shown in *Figure 3.1* or select "Layers . . .?" from the Format pull down menu.

Either method will call the Layer and Linetype Properties dialog box illustrated in *Figure 3.2.* This is a tabbed "index card" style box, like the Help topics dialog presented in the last chapter. There are only two tabs, Layers and Linetypes. Yours should be showing the Layers card. The large open rectangle in the middle shows the names and properties of all layers defined in the current drawing. Layering systems can become very large and complex and for this reason there is a system to limit the layer names shown in the name window. This is controlled by the box at the top left which currently says "All", meaning all layers are shown. To the right of this box, the current layer is identified. Once we have defined new layers we will be able to use the Current button to change the current layer.

In the layer name list window, you will see that "0" is the only defined layer. The icons on the line after the layer name show the current state of various properties that the layer. We will get to this shortly..

Now we will create three new layers. In Release 14 this is extremely easy.

⌗ Click on "New", to the right of the list window.

The newly defined layer, layer 1, will be entered immediately in the layer name window. You will see that it is currently defined with characteristics identical to layer 0. We will alter these momentarily. But first give this layer a name and then define two more layers.

Figure 3.1.

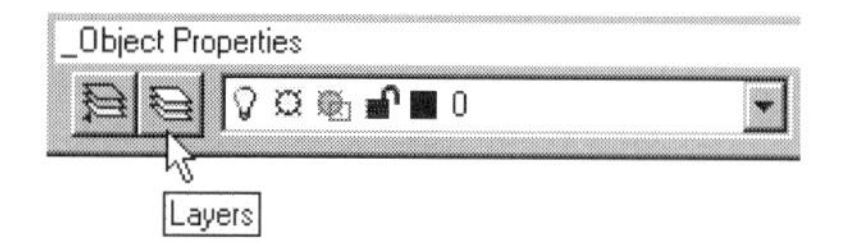

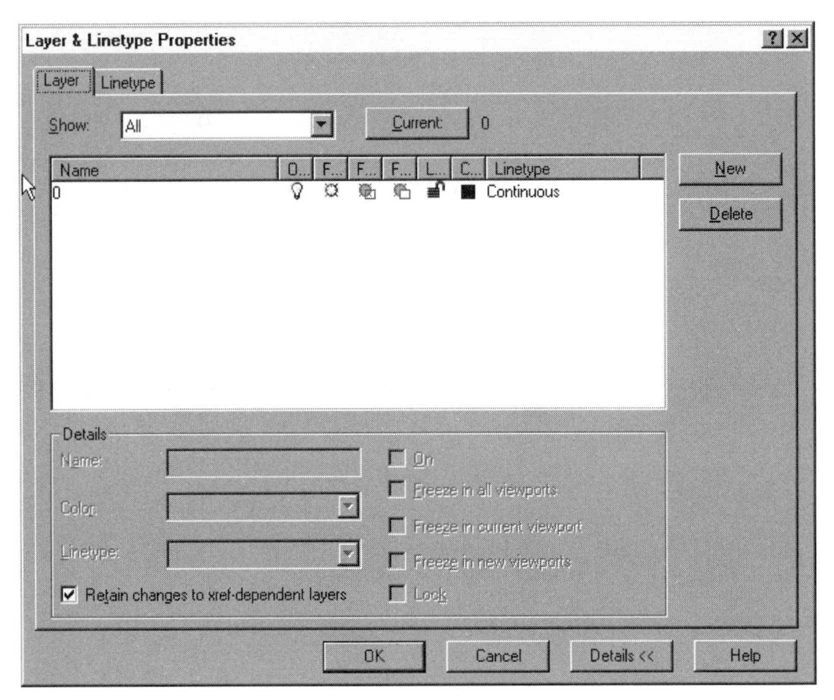

Figure 3.2.

Layer names may be up to 31 characters long. We have chosen single-digit numbers as layer names because they are easy to type, they are the AutoCAD default names, and we can match them to AutoCAD's color numbering sequence.

⌖ **Type "1" for the layer name.**

"Layer1" will change to simply "1".

⌖ **Click on "New" again.**

A second new layer, will be added to the list. It will again have the default name "Layer1". Change it to "2"

⌖ **Type "2" for the second layer name.**

⌖ **Click on "New" once more.**

⌖ **Type "3" for the third layer name.**

At this point your layer name list should show layers 0,1,2, and 3, all with identical properties.

3.2 ASSIGNING COLORS TO LAYERS

We now have four layers, but they are all pretty much the same. Obviously we have more changes to make before our new layers will have useful identities.

It is common practice to leave layer 0 defined the way it is. We will begin our changes on layer 1.

⌖ **(If for any reason you have left the Layer and Linetype Properties dialog box, reenter it by selecting the Layers tool from the Object properties toolbar or "Layers . . ." from the Format menu.)**

Before you can change the qualities of a layer, you must select it in the layer name box.

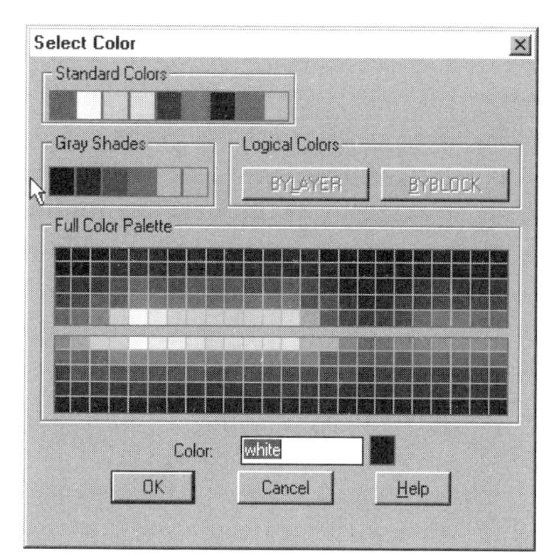

Figure 3.3.

⌗ **Move the cursor arrow anywhere on the layer 1 line and click to select layer1.**

The layer name will be highlighted.

⌗ **Click on the white square under "C".**

The C is for color. The columns may be made wider, if you wish so that the complete heading is visible. Clicking on the white square in the color column will call the Select Color dialog box illustrated in *Figure 3.3.* There are nine standard colors at the top, followed by gray shades and a palette of colors.

AutoCAD can display up to 256 different color shades. The 9 standard colors shown at the top of the box are numbered 1 through 9 and are the same for all color monitors. Color numbers 10–249 are shown in the full color palette and numbers 250–255 are the gray shades shown in between.

⌗ **Select the red box, the first of the nine standard colors at the top of the box.**

⌗ **Click on "OK".**

You will now see that layer 1 is defined with the color red in the layer name list box.

Next we will assign the color yellow to layer 2.

⌗ **Click on the layer 2 line.**

2 should now be highlighted.

⌗ **Click on the white square under Color and assign the color yellow (color #2) to layer 2 in the Select Color dialog box.**

⌗ **Click on OK.**

⌗ **Select layer 3 and set this layer to green, color #3.**

Look at the layer list. You should now have the layers 1,2, and 3 defined with the colors red, yellow and green.

3.3 ASSIGNING LINETYPES

AutoCAD has a standard library of linetypes that can be assigned easily to layers. There are 45 standard types in addition to continuous lines. If you do not assign a linetype, AutoCAD will assume you want continuous lines. In addition to continuous lines we will be using hidden and center lines. We will put hidden lines in yellow on layer 2 and center lines in green on layer 3.

⌖ (If for any reason you have left the Layer and Linetype dialog box, reenter it by selecting the Layer tool from the Object Properties toolbar.)

⌖ Click on "Continuous" in the Linetype column of the layer 2 line.

This will select layer 2 and call up the Select Linetype dialog box illustrated in *Figure 3.4* The box containing a list of loaded linetypes currently only shows the continuous linetype. We can fix this by selecting the "Load . . ." button at the bottom of the dialog box.

NOTE: Make sure that you actually click on the word "continuous". If you click on one of the icons you may turn a layer off or freeze it so that you cannot draw on it.

⌖ Click on "Load".

This will call a second dialog box, the Load or Reload Linetypes box illustrated in *Figure 3.5*. Here you can pick from the list of linetypes available from the standard "acad" file or from other files containing linetypes, if there are any on your system. You also have the option of loading all linetypes from any given file at once. The linetypes are then defined in your drawing and you can assign a new linetype to a layer at any time. This makes things easier. It does, however, use up more memory.

For our purposes we will load only the hidden and center linetypes we are going to be using.

⌖ Scroll down until you see the Center linetype.

⌖ Click on "Center" in the Linetype column at the left.

⌖ Scroll down again until you see the "Hidden" linetype on the list.

⌖ Hold down the Ctrl key and click "Hidden" in the Linetype column.

The Ctrl key lets you highlight two separated items in a list.

⌖ Click on "OK" to complete the loading process.

You should now see the center and hidden linetypes added to the list of loaded linetypes. Now that these are loaded, we can assign them to layers.

⌖ Click on "Hidden" in the linetype column.

⌖ Click on "OK" to close the box.

You should see that layer 2 now has the hidden linetype.

Figure 3.4.

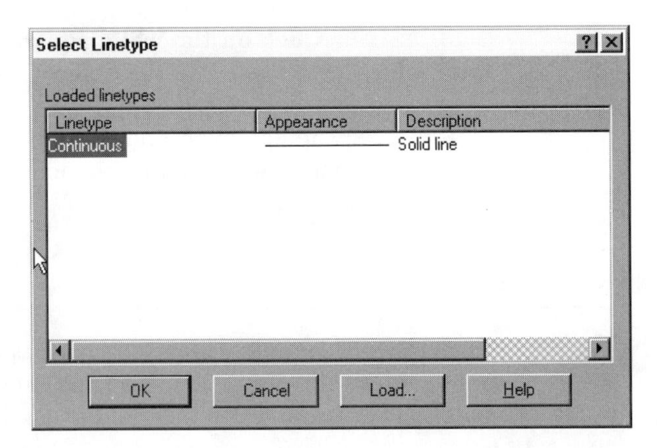

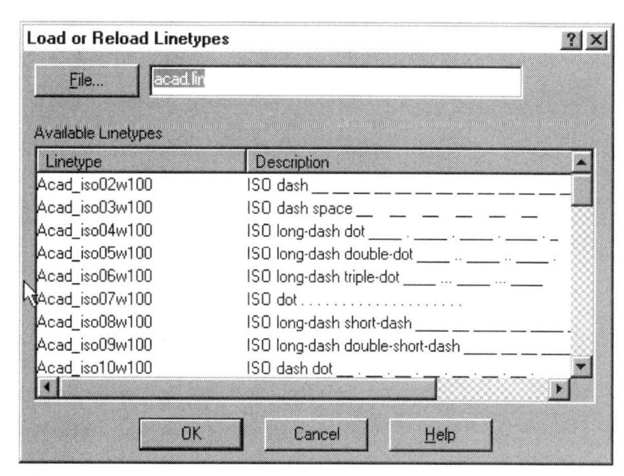

Figure 3.5.

Next assign the center linetype to layer 3.

⌖ **Click on "Continuous" in the linetype column of the layer 3 line.**

⌖ **In the Select Linetype dialog box select the Center linetype.**

⌖ **Click on OK.**

Examine your layer list again. It should show layer 2 with the hidden linetype and layer 3 with the center linetype.

⌖ **Click on OK to exit the Layer and Linetype dialog box.**

3.4 CHANGING THE CURRENT LAYER

In order to draw new entities on a layer, you must make it the currently active layer. Previously drawn objects on other layers that are turned on will be visible also and will be plotted, but new objects will go on the current layer.

You can make a layer current through the Layer and Linetype dialog box, but a quicker method is to use the layer list on the Object Properties toolbar.

⌖ **Click anywhere in the current layer list box on the Object Properties toolbar.**

This will open the list, as shown in *Figure 3.6.*

⌖ **Select Layer1.**

Layer 1 will replace layer 0 as the current layer on the Object Properties toolbar. At this point we suggest that you try drawing some lines to see that you are, in fact, on layer 1 and drawing in red, continuous lines.

When you are satisfied with the red lines you have drawn, leave them on the screen and switch to layer 2.

Figure 3.6.

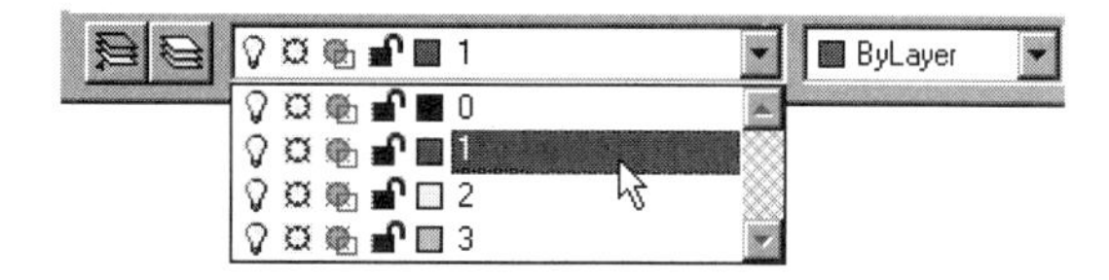

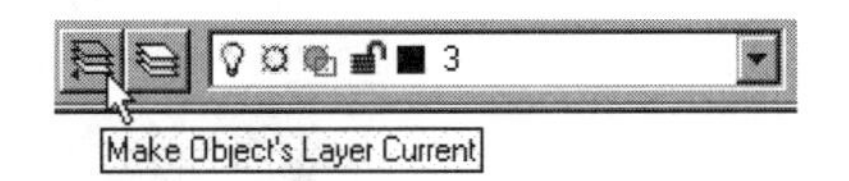

Figure 3.7.

⌗ **Click anywhere in the Layer Control box on the Object Properties toolbar.**

⌗ **Click on the layer 2 line.**

Layer 2 will become the current layer

Now draw more lines and see that they are "hidden" yellow lines.

⌗ **Make layer 3 the current layer and draw some green center lines.**

Finally, we will use another method to make layer 1 current before moving on.

⌗ **Select the Make Object's Layer Current tool from the Object Properties toolbar, as shown in *Figure 3.7.***

This tool allows us to make a layer current by selecting any object on that layer. AutoCAD will show this prompt:

Select object whose layer will become current:

⌗ **Select any red, continuous line, drawn on layer 1.**

Layer 1 will replace layer 3 in the current layer box.

3.5 EDITING CORNERS USING FILLET

Now that you have a variety of linetypes to use, you can begin to do some more realistic mechanical drawings. All you will need is the ability to create filleted (rounded) and chamfered (cut) corners. The two work similarly, and AutoCAD makes them easy. Fillets may also be created between circles and arcs, but the most common usage is the type of situation demonstrated here.

⌗ **Erase any lines left on the screen from previous exercises.**

⌗ **If you have not already done so, set Layerl as the current layer.**

⌗ **Draw a 5 x 5 square on your screen, as in *Figure 3.8.***

We will use this figure to practice fillets and chamfers. Exact coordinates and lengths are not significant.

Figure 3.8.

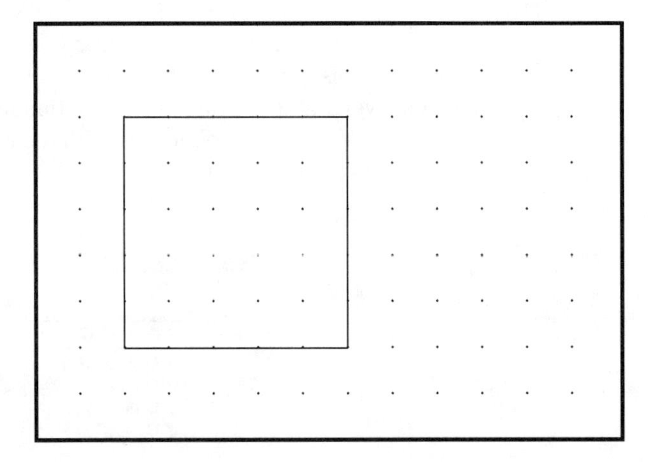

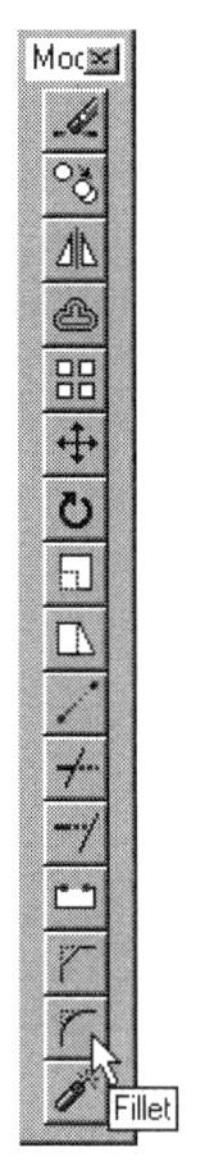

Figure 3.9.

⌗ **Type "f", select the Fillet tool from the Modify toolbar, or select Fillet from the Modify menu.**

Depending on your monitor, you may have to move the toolbar from its docked position in order to see the Fillet tool, which is at the bottom of the Modify toolbar. Undock the toolbar by clicking in the gray area and dragging the toolbar to themiddle of the screen. Once the toolbar is floating, you will be able to select the fillet tool, as shown in Figure 3.9. Moving the toolbar is cumbersome, however, so if you do have to do this, typing "f" or selecting from the Modify menu is preferable.

A prompt with options will appear as follows:

(TRIM mode) Current fillet radius = 0.50
Polyline/Radius/Trim<Select first object>:

The first thing you must do is determine the degree of rounding you want. Since fillets are really arcs, they can be defined by a radius.

⌗ **Type "r".**

AutoCAD prompts:

Enter fillet radius <0.50>:

The default is 0.50.

⌗ **Type ".75" or show two points .75 units apart.**

You have set .75 as the current fillet radius for this drawing. You can change it at any time. Changing will not affect previously drawn fillets.

⌗ **Press Enter or the space bar to repeat FILLET.**

The prompt is the same as before, but the radius value has changed:

(TRIM mode) Current fillet radius = 0.75
Polyline/Radius/Trim<Select first object>:

You will notice that you have the pickbox on the screen now without the cross hairs.

⌗ **Use the pickbox to select two lines that meet at any corner of your square.**

Behold! A fillet! You did not even have to press Enter. AutoCAD knows that you are done after selecting two lines.

⌗ **Press Enter or the space bar to repeat FILLET. Then fillet another corner.**

We suggest that you proceed to fillet all four corners of the square. When you are done your screen should resemble *Figure 3.10*.

Figure 3.10.

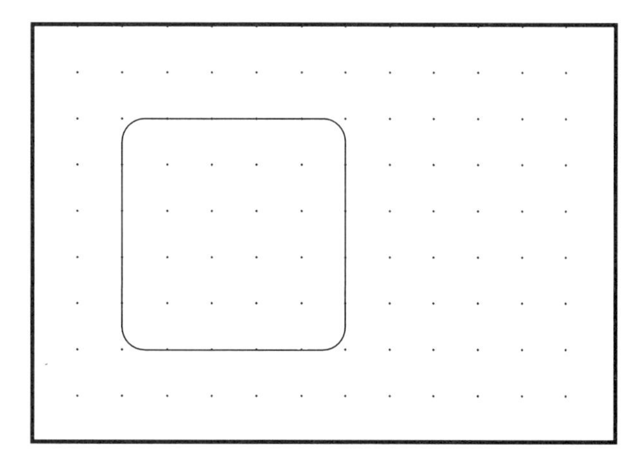

3.6 EDITING CORNERS WITH CHAMFER

The CHAMFER command sequence is almost identical to the FILLET command, with the exception that chamfers may be uneven. That is, you may cut back farther on one side of a corner than on the other. To do this you must give AutoCAD two distances instead of one.

⊕ **Prepare for this exercise by undoing all your fillets using the U command as many times as necessary.**

⊕ **Type "cha" or select Chamfer from the Modify menu.**

(Here again, you can may have to have the Modify toolbar in a floating position in order to select the Chamfer tool, shown previously in *Figure 3.9.*)

AutoCAD prompts:

(TRIM mode) Current chamfer Dist1 = 0.50, Dist2 = 0.50
Polyline/Distance/Angle/Trim/Method/<Select first line>:

⊕ **Type "d".**

The next prompt will be:

Enter first chamfer distance <0.50>:

⊕ **Type "1".**

AutoCAD asks for another distance:

Enter second chamfer distance <1.00>:

The first distance has become the default and most of the time it will be used. If you want an asymmetric chamfer, enter a different value for the second distance.

⊕ **Press enter to accept the default, making the chamfer distances symmetrical.**

⊕ **Press Enter to repeat the CHAMFER command.**

⊕ **Answer the prompt by pointing to a line this time.**

⊕ **Point to a second line, perpendicular to the first.**

You should now have a neat chamfer on your square. We suggest that you continue this exercise by chamfering the other three corners of your square.

3.7 ZOOMING WINDOW, PREVIOUS, AND ALL

The capacity to zoom in and out of a drawing is one of the more impressive benefits of working on a CAD system. When drawings get complex it often becomes necessary to work in detail on small portions of the drawing space. Especially with a small monitor, the only way to do this is by making the detailed area larger on the screen. This is easily done with the ZOOM command.

⊕ **You should have a square with chamfered corners on your screen from the previous exercise.**

We will demonstrate zooming using the window, all, previous, and realtime options.

Figure 3.11.

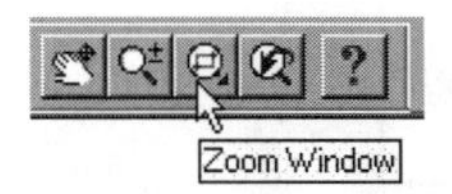

⊕ **Type "z" or select the Zoom Window tool from the Standard toolbar, as illustrated in *Figure 3.11.***

The prompt that follows includes the following options:

All/Center/Dynamic/Extents/Previous/Scale(X/XP)/ Window/<Realtime>:

If you have used the Zoom Window tool, the Window option will be entered automatically. As in ERASE and other edit commands, you can force a window selection by typing "w" or selecting "Window". However, this is unnecessary. The windowing action is automatically initiated if you pick a point on the screen after entering ZOOM.

⌗ **Pick a point just below and to the left of the lower left-hand corner of your square (point 1 in *Figure 3.12*).**

AutoCAD asks for another point:

Other corner:

You are being asked to define a window, just as in the ERASE command. This window will be the basis for what AutoCAD displays next. Since you are not going to make a window that exactly conforms to the screen size and shape, AutoCAD will interpret the window this way: Everything in the window will be shown, plus whatever additional area is needed to fill the screen. The center of the window will become the center of the new display.

⌗ **Pick a second point near the center of your square (point 2 in the figure).**

The lower left corner of the square should now appear enlarged on your screen, as shown in *Figure 3.13.*

⌗ **Using the same method, try zooming up further on the chamfered corner of the square. If snap is on you may need to turn it off (F9).**

Remember that you can repeat the ZOOM command by pressing Enter or the space bar.

At this point, most people cannot resist seeing how much magnification they can get by zooming repeatedly on the same corner or angle of a chamfer. Go ahead. After a couple of zooms the angle will not appear to change. An angle is the same angle no matter how close you get to it. But what happens to the spacing of the grid and snap as you move in?

When you are through experimenting with window zooming, try zooming to the previous display.

Figure 3.12.

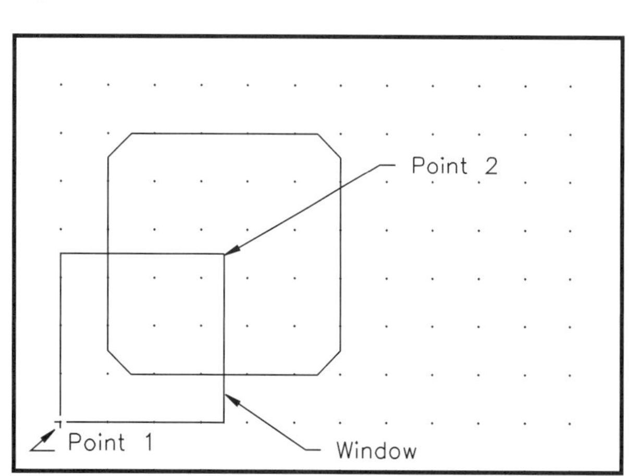

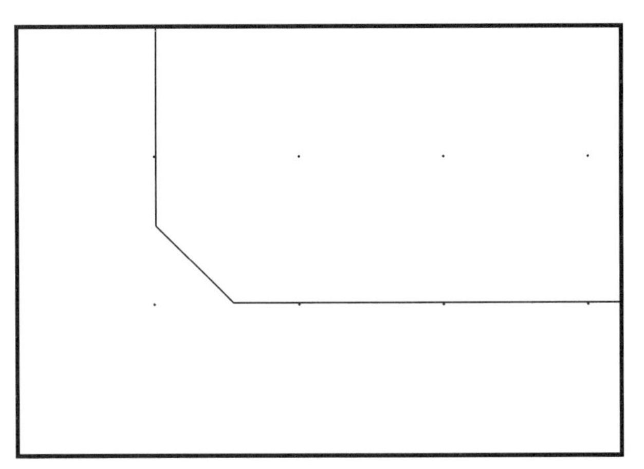

Figure 3.13.

Zoom Previous

⌖ Press Enter to repeat the ZOOM command, or select the Zoom Previous tool, as shown in *Figure 3.14.*

⌖ If you repeated the command by pressing enter, type "p".

You should now see your previous display.

AutoCAD keeps track of up to ten previous displays.

⌖ ZOOM "Previous" as many times as you can until you get a message that says:

No previous display saved.

Zoom All

ZOOM All zooms out to display the whole drawing. It is useful when you have been working in a number of small areas of a drawing and are ready to view the whole scene. You do not want to have to wade through previous displays to find your way back. ZOOM All will take you there in one jump.

In order to see it work, you should be zoomed in on a portion of your display before executing ZOOM All.

⌖ Press Enter or type "z" to repeat the ZOOM command and zoom in on a window within your drawing.

⌖ Press Enter or type type "z" to repeat ZOOM again.

⌖ Type "a" for the All option.

Figure 3.15.

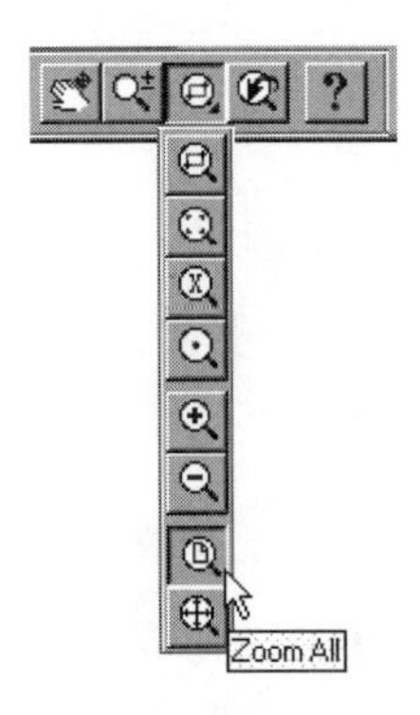

NOTE: The Zoom Window tool also has a flyout that includes tools for all of the ZOOM command options, including Zoom All, as shown in *Figure 3.15.* Flyouts are toolbar features that make additional tools available. Any tool button that has a small black triangle in one corner will open a flyout. To open a flyout hold down the pick button while

Figure 3.14.

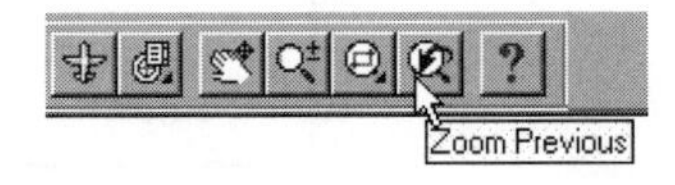

the arrow is on the tool. Then run down or across the flyout to the tool you want. When the tool button is "down" release the pick button.

3.8 REALTIME ZOOM AND PAN

Release 14 has new Realtime ZOOM and PAN features. Realtime means that you will see changes in display and magnification dynamically as you make adjustments. As soon as you start to use ZOOM you are likely to need PAN as well. While ZOOM allows you to magnify portions of your drawing, PAN allows you to shift the area you are viewing in any direction.

In Windows you have the option of panning with the scrollbars at the edges of the drawing area, but the PAN command allows more precision and flexibility. In particular it allows diagonal motion. We will use the PAN Realtime tool first and then the ZOOM Realtime tool.

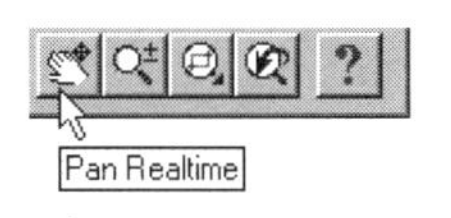

Figure 3.16.

Type "p" or select the Pan tool from the Standard toolbar (as illustrated in *Figure 3.16*).

With either method, you will see the PAN command cursor representing a hand with which you can move objects on the screen. When the mouse button is not depressed, the hand will move freely across the screen. When you move the mouse with the pick button held down, the complete drawing display will move along with the hand.

Move the hand near the middle of the screen without holding down the pick button.

Hold down the pick button and move the cursor diagonally up and to the right, as illustrated in *Figure 3.17*.

Release the pick button to complete the PAN procedure.

Experiment with Realtime PAN, moving objects up, down, left, right, and diagonally. You should also compare the effect of using the scrollbars with the PAN command.

Realtime ZOOM

Any time that you are in either the ZOOM or PAN command, you can access a pop up menu by pressing the right button on your cursor. Try it.

Figure 3.17.

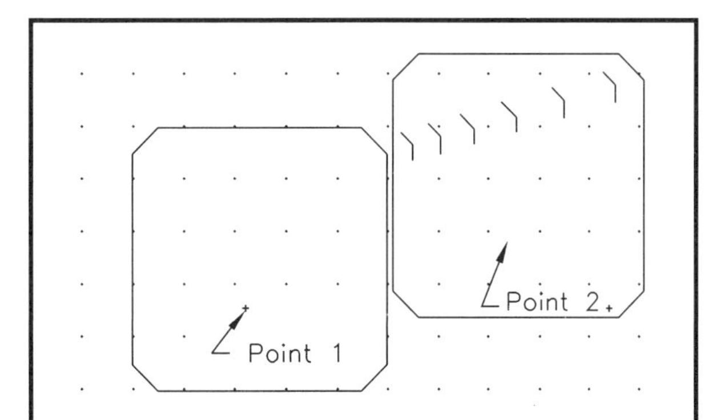

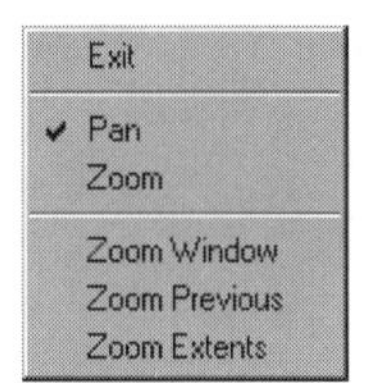

Figure 3.18.

Without leaving PAN, press the right button on your mouse.

This will call up the pop up menu illustrated in Figure 3.18. Pop up lists are context specific. They are called only from within certain commands and contain options that are useful in specific conditions. This menu is useful while you are panning and zooming. In particular, it will allow you to quickly switch between realtime PAN and realtime ZOOM.

Select Zoom from the pop up menu.

The menu will vanish and the realtime ZOOM cursor will appear. As illustrated in *Figure 3.19,* this represents a magnifying glass with a + sign above and a – sign below.

Figure 3.19.

Without pressing the pick button, move the zoom cursor near the bottom of the screen.

As with the PAN cursor, you will be able to move freely when the pick button is not held down.

Press the pick button down and move the cursor upwards.

With the pick button down, upward motion will increase magnification, enlarging objects on the screen.

Move up and down to see the effects of zoom cursor movment.

Release the pick button to complete the process. When you release the button you will not exit the command. This is important because it may take several trips up or down the screen to indicate the amount of magnification you want. Moving the cursor halfway up the screen will produce a 100% magification.

Continue to experiment with Realtime ZOOM and PAN until you feel comfortable. Then exit the command.

To exit, press ESC, the space bar, or the enter key.

You can also press the right button to open the pop-up menu, and then select Exit.

3.9 USING PLOT PREVIEW

The plot preview feature of the Plot Configuration dialog box is an essential tool in carrying out efficient plotting and printing. Plot configuration is complex and the odds are good that you will waste time and paper by printing drawings directly without first previewing them on the screen. In this task we are still significantly limited in our use of plot configuration parameters, but learning to use the plot preview will make all of your future work with plotting and printing more effective.

As in Chapter 2, we suggest you work through this task now with whatever objects are on your screen, and return to it after you have done one of the drawings at the end of the chapter.

To begin this task you should have objects or a drawing on your screen ready to preview.

Any drawing will do, but we will illustrate using Drawing 3—2.

Type CTRL-P, or select the Print tool from the Standard toolbar or Print from the File menu.

This will open the Plot Configuration dialog box. Plot Preview is at the lower right of the dialog box. There are two types of preview, "Partial" and "Full". A partial preview will show you an outline of the effective plotting area in relation to the paper size, but will not show an image of the plotted drawing.

NOTE: We will address paper sizes in the next chapter. For now, we will assume that your plot configuration is correctly matched to the paper in your printer or plotter. If you do not get good results with this task, the problem may very well lie here.

Partial Preview

Partial previews are quick and should be accessed frequently as you set plot parameters that affect how paper will be used and oriented to print your drawing. Full previews take longer and may be saved for when you think you have got everything right.

We will look at a partial preview first.

⌖ **If the "Partial" radio button is not selected in your dialog box, select it now.**

⌖ **Click on "Preview".**

You will see a preview image similar to the one shown in *Figure 3.20.* The exact image will depend on your plotting device, so it may be different from the one shown here. The elements of the preview will be the same, however. The red rectangle represents your drawing paper. It may be oriented in landscape (horizontally) or portrait (vertically). Typically, printers are in portrait, and plotters in landscape.

The blue rectangle inside the red one illustrates the effective plotting area. It is quite possible that the blue and red rectangles will overlap. The blue represents the size and shape of the area that AutoCAD can actually use given the shape of the drawing in relation to the size and orientation of the paper. The effective area is dependent on many things, as you will see. We will leave it as is for now and return later when you begin to alter the plot configuration.

In one corner of the red rectangle you will see a small triangle. This is the rotation icon. It shows the corner of the plotting area where the plotter will begin plotting (the origin).

⌖ **Click on "OK" to exit the preview box.**

Full Preview

⌖ **Click on the "Full" radio button to switch to a full preview.**

⌖ **Click on "Preview".**

Figure 3.20.

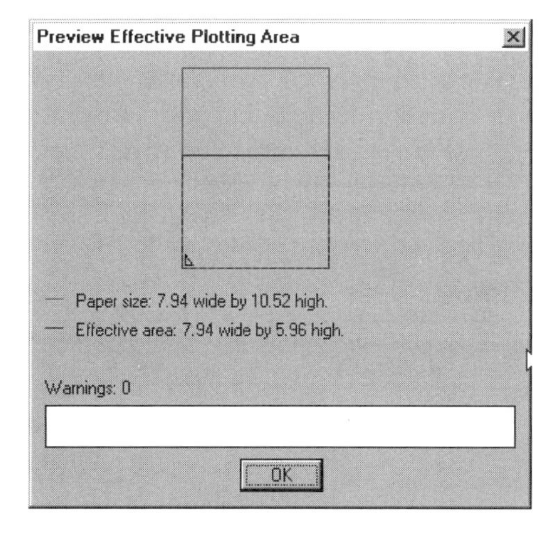

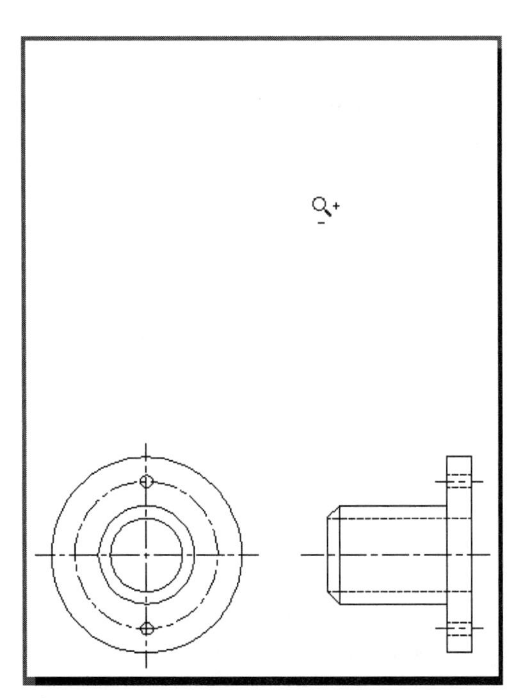

Figure 3.21.

The dialog box will disappear temporarily and you will see a preview similar to the one in *Figure 3.21*. We have used Drawing 3—2 to illustrate. You will see whatever drawing you are previewing, with orientation and placement depending on your plotting device.

This image represents your drawing on paper as it is now configured for printing. The ZOOM Realtime cursor appears to allow you to zoom in or out on aspects of the preview. By clicking the right button you can access the zoom and pan pop-up list demonstrated earlier in this chapter. Panning and zooming in the preview has no effect on the plot parameters. You may want to experiment with this feature now.

When you are done experimenting, press ESC or the enter key or space bar to return to the Plot Configuration dialog box.

This ends our initial preview of your drawing. In ordinary practice, if everything looked right in the preview you would move on to plot or print your drawing now by clicking on "OK". In the chapters that follow, we will explore the rest of the plot configuration dialog box and will use plot preview extensively as we change plot parameters. For now, get used to using plot preview. If things are not coming out quite the way you want, you will be able to fix them soon.

3.10 REVIEW MATERIAL

Questions

1. What function(s) can be performed directly from the Layer list on the Object Properties toolbar? What functions can be performed from the Layer and Linetype dialog box?

2. What linetype is always available when you start a drawing from scratch in AutoCAD? What must you do to access other linetypes?
3. How many colors are available in AutoCAD?
4. How many different layers does AutoCAD allow you to create?
5. Name three ways to change the current layer.
6. Why is it often necessary to enter the FILLET and CHAMFER commands twice in order to create a single fillet or chamfer?
7. What happens to the grid when you zoom way out on a drawing?
8. Name one limitation of the scroll bars that the PAN command does not have.
9. What is the difference between a partial and a full plot preview?

COMMANDS

Data	View	Modify
LAYER	ZOOM	CHAMFER
LTSCALE	PAN	FILLET
	REGEN	

Drawing Problems

1. Make layer 3 current and draw a green center line cross with two perpendicular lines, each 2 units long and intersecting at their midpoints.
2. Make layer 2 current and draw a hidden line circle centered at the intersection of the cross drawn in problem 1, with radius of 2 units.
3. Make layer 1 current and draw a red square 2 units on a side centered on the center of the circle. Its sides will run tangent to the circle.
4. Use a window to zoom in on the objects drawn in problems 1, 2 and 3.
5. Fillet each corner of the square with a .125 radius fillet.

DRAWING 3—1

Mounting Plate

This drawing will give you experience using center lines and chamfers. Since there are no hidden lines, you will have no need for layer 2, but we will continue to use the same numbering system for consistency. Draw the continuous lines in red on layer 1 and the center lines in green on layer 3.

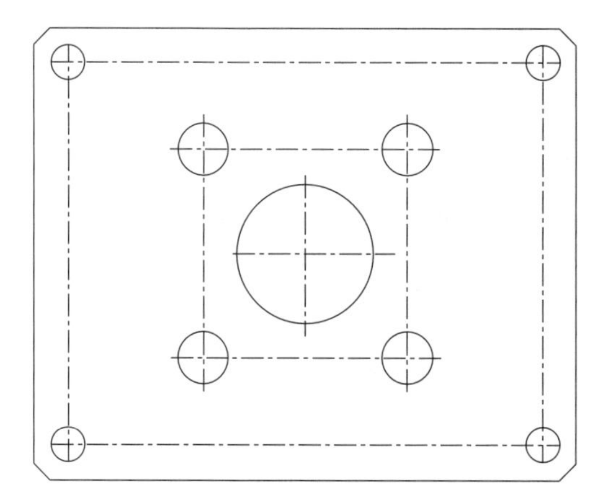

Drawing Suggestions

GRID = .5
SNAP = .25
LTSCALE = .5

Ltscale The size of the individual dashes and spaces that make up center lines, hidden lines, and other linetypes is determined by a global setting called "LTSCALE". By default it is set to a factor of 1.00. In smaller drawings this setting will be too large and will cause some of the shorter lines to appear continuous regardless of what layer they are on.

To remedy this, change LTSCALE as follows:

1. Type "ltscale" or select "Linetypes" and then "Global Linetype Scale" from the Options pull down menu.
2. Enter a value.

For the drawings in this chapter use a setting of .50. See *Figure 3.22* for some examples of the effect of changing LTSCALE.

NOTE: In manual drafting it would be more common to draw the center lines first and use them to position the circles. Either order is fine, but be aware that what is standard practice in pencil and paper drafting may not be efficient or necessary on a CAD system.

LTSCALE = 1.00

LTSCALE = .50

LTSCALE = .25

Figure 3.22.

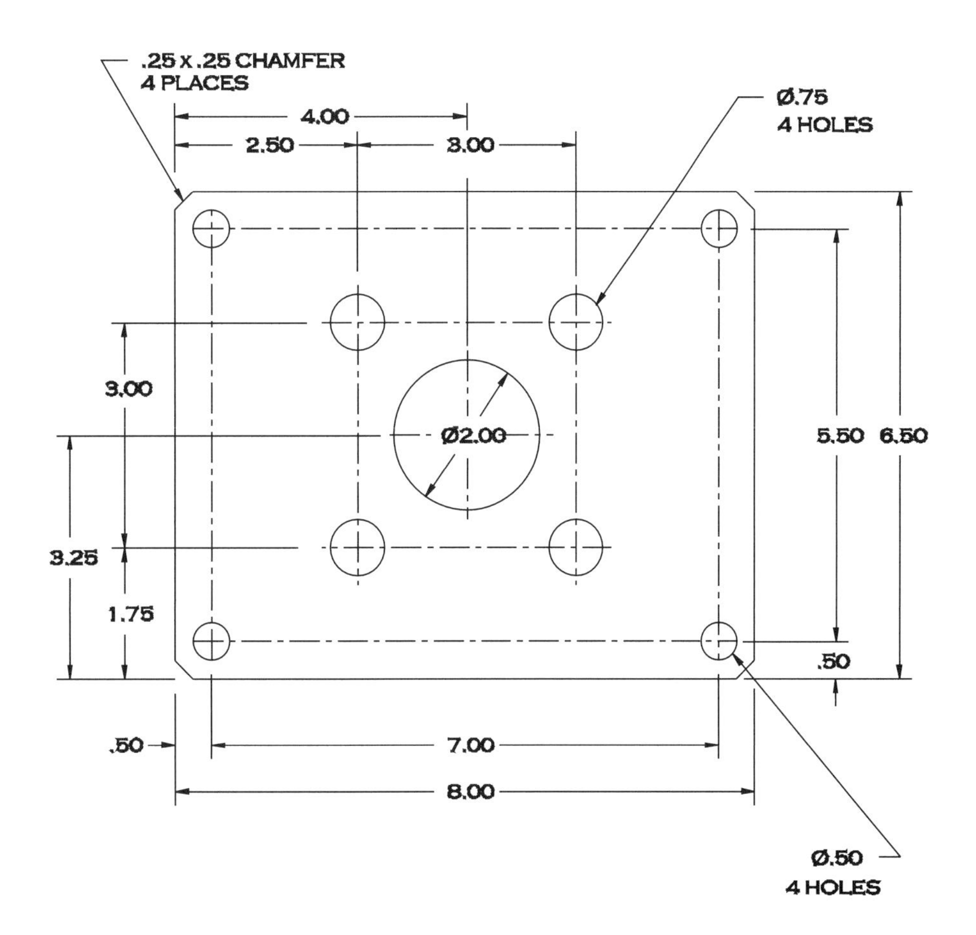

MOUNTING PLATE

DRAWING 3-1

DRAWING 3—2

BUSHING

This drawing will give you practice with chamfers, layers, and zooming. Notice that because of the smaller dimensions here, we have recommended a smaller LTSCALE setting.

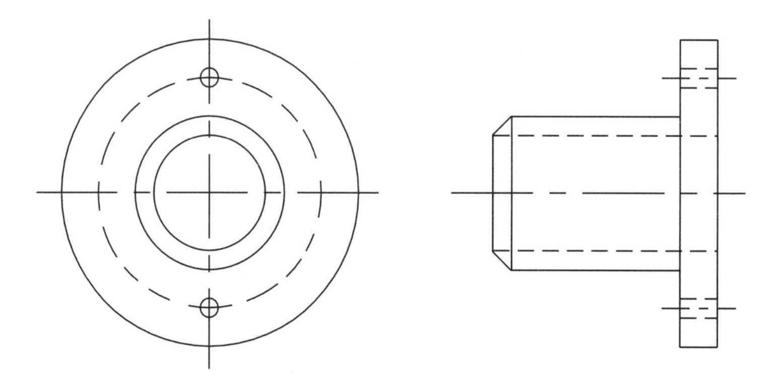

Drawing Suggestions

GRID = .25
SNAP = .125
LTSCALE = .25

- Since this drawing will appear quite small on your screen, it would be a good idea to ZOOM in on the actual drawing space you are using, and use PAN if necessary.
- Notice that the two .25-diameter screw holes are 1.50 apart. This puts them squarely on grid points that you will have no trouble finding.

Regen When you zoom you may find that your circles turn into many-sided polygons. AutoCAD does this to save time. These time savings are not noticeable now, but when you get into larger drawings they become very significant. If you want to see a proper circle, type "Regen". This command will cause your drawing to be regenerated more precisely from the data you have given.

You may also notice that REGENs happen automatically when certain operations are performed, such as changing the LTSCALE setting after objects are already on the screen.

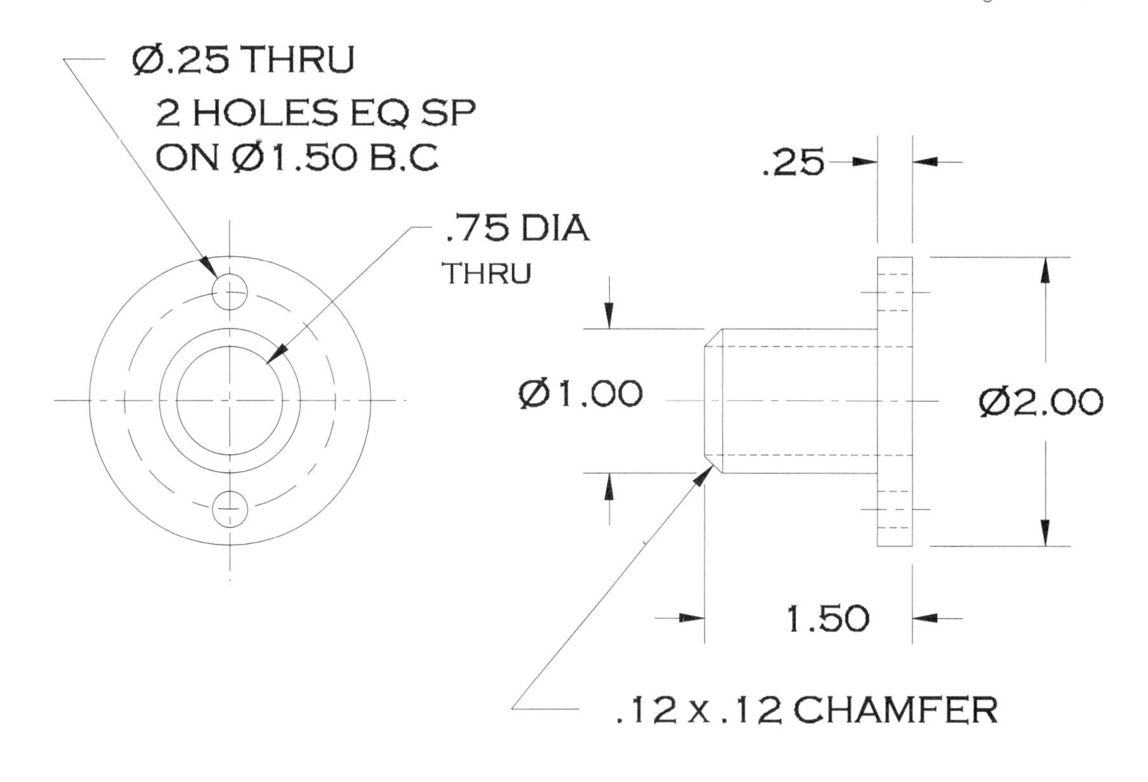

BUSHING
DRAWING 3-2

DRAWING 3—3

Half Block

This cinder block is the first project using architectural units in this book. Set units, grid, and snap as indicated, and everything will fall into place nicely.

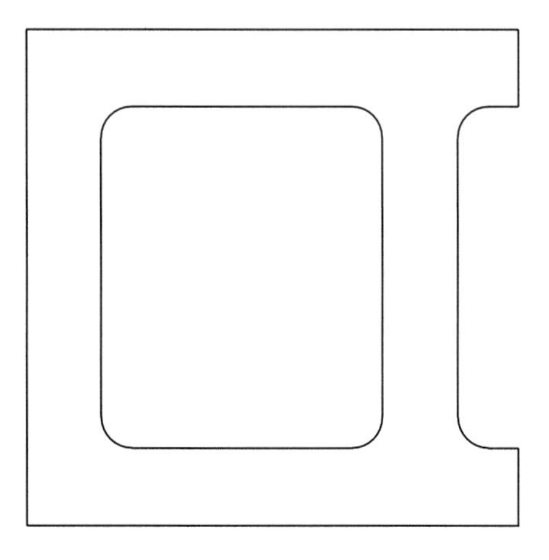

Drawing Suggestions

UNITS = Architectural smallest fraction = 4(1/4″)
GRID = 1/4″
SNAP = 1/4″

- Start with the lower left corner of the block at the point (0′-1″,0′-1″) to keep the drawing well placed on the display.
- After drawing the outside of the block with the 5 1/2″ indentation on the right, use the DIST command to locate the inner rectangle 1 1/4″ in from each side.
- Set the FILLET radius to 1/2″ or .5. Notice that you can use decimal versions of fractions. The advantage is that they are easier to type.

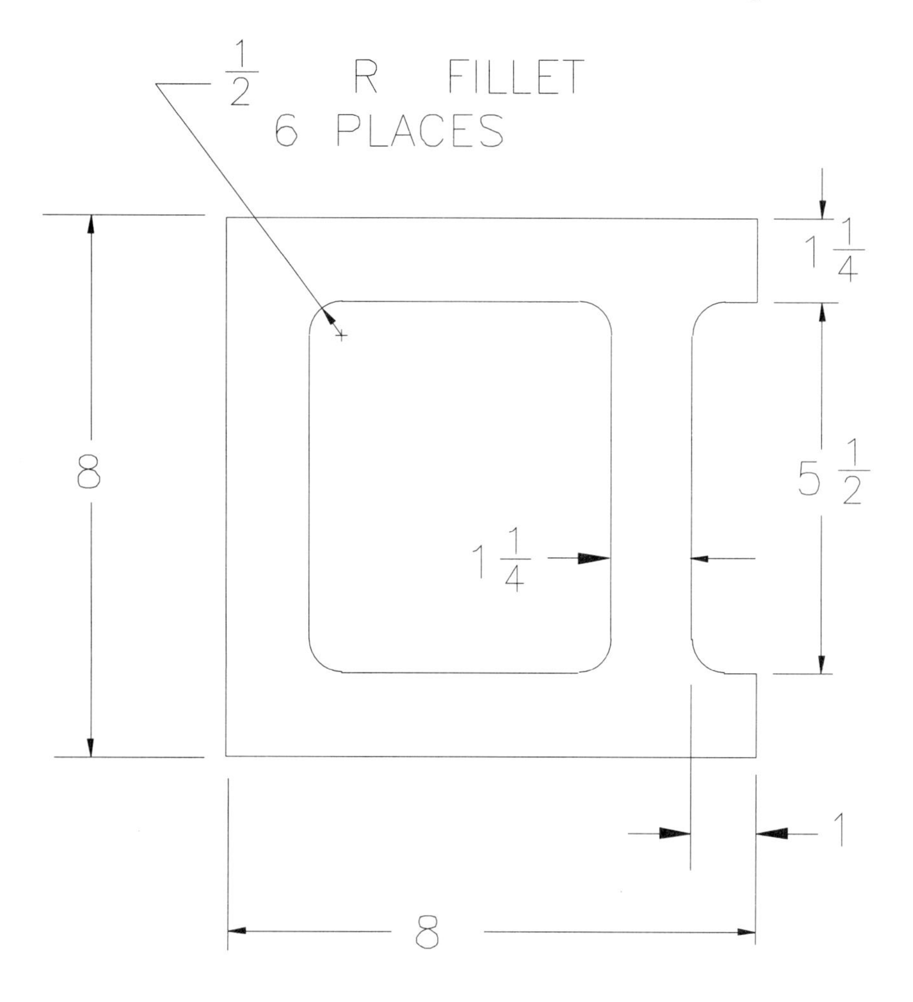

HALF BLOCK

Drawing 3–3

4

Template Drawings

Overview

In this chapter you will learn some real timesavers. If you have grown tired of defining the same three layers, along with units, grid, snap, and ltscale, for each new drawing, read on. You are about to learn how to use template drawings. With templates, you can begin each new drawing with whatever setup you want.

In addition, you will learn to reshape the grid using the LIMITS command, and to COPY, MOVE, and ARRAY objects on the screen so that you do not have to draw the same thing twice. We will begin with LIMITS, since we will want to change the limits as part of defining your first template.

Sections

Objectives

After reading this chapter, you should be able to:

- Change the shape of the grid using the Quick Setup Wizard and the LIMITS command.
- Create a template drawing.
- Select your drawing as the template.
- MOVE an object in a drawing.
- COPY an object in a drawing.
- Create a rectangular ARRAY.
- Change Plot Configuration parameters.
- Draw a grill.
- Draw a test bracket.
- Draw a floor frame.

4.1 SETTING LIMITS

You have changed the density of the screen grid many times, but always within the same 12 × 9 space, which basically represents an A-size sheet of paper. Now you will learn how to change the shape, by setting new limits to emulate other sheet sizes or any other space you want to represent. But first, a word about model space and paper space.

Model Space and Paper Space

"Model space" is an AutoCAD concept that refers to the imaginary space in which we create and edit objects. In model space, objects are always drawn full scale (1 screen unit = 1 unit of length in the real world). The alternative to model space is "paper space", in which screen units represent units of length on a piece of drawing paper. Paper space is most useful in plotting multiple views of 3D drawings. In Part I of this book all of your work will be done in model space.

In this exercise we will reshape our model space to emulate different drawing sheet sizes. This will not be necessary in actual practice. With AutoCAD you can scale your drawing to fit any drawing sheet size when it comes time to plot. Model space limits should be determined by the size and shapes of objects in your drawing, not by the paper you are going to use when you plot.

The Quick Setup Wizard

Limits can be set from within a drawing using the LIMITS command. When you are beginning a new drawing, however, a quicker way is to use a drawing setup wizard. Wizards are standard Windows 95 features that guide you through setup functions in all kinds of applications. We will use the Quick Setup Wizard to begin this drawing and set 18 × 12 limits.

Start AutoCAD or begin a new drawing.

If the Create New Drawing dialog box is not showing, open the File menu and select New, or select the New tool from the Standard toolbar.

In the Create New Drawing dialog box click the Use a Wizard button on the left.

Highlight Quick Setup in the Select a Wizard box.

Click OK.

This will bring up the Quick Setup tabbed dialog box illustrated in *Figure 4.1*. Our illustration shows the Step 2 Area card on top because this is where you will set new limits. The Step 1 Units card will allow you to select decimal units, the default, but will not allow you to set the number of decimal places to use. For this you will need to continue using the UNITS command for now.

Figure 4.1.

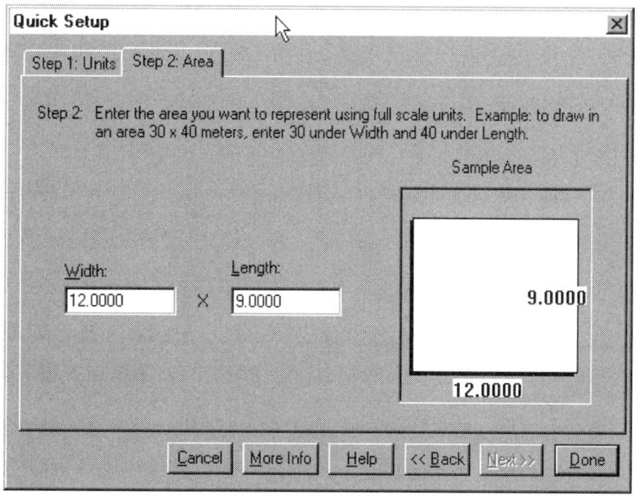

- **Click on the Step 2 Area card to bring it to the top.**
- **Double click in the Width box to highlight the number 12.0000.**
- **Type "18".**

The number in the width box should now read 18.0000. Also the number at the bottom of the preview image on the right will change to 18.0000.

- **Double click in the Length box and type "12".**

The length box and the preview image will show 12.0000.

- **Click on Done to exit the Quick Setup Wizard.**

Your screen should resemble *Figure 4.2,* showing an 18 × 12 grid.

After we are finished exploring the LIMITS command, we will create the new settings we want and save this drawing as your B-size template.

We suggest that you continue to experiment with setting limits, and that you try some of the possibilities listed in *Figure 4.3,* which is a table of sheet sizes. Now that you are already in a drawing, you may wish to use the LIMITS command to try other settings.

The LIMITS command

The LIMITS command changes the grid area in the same way as the Quick Setup Wizard, except that it is used after a drawing has already been opened and uses coodinate

Figure 4.2.

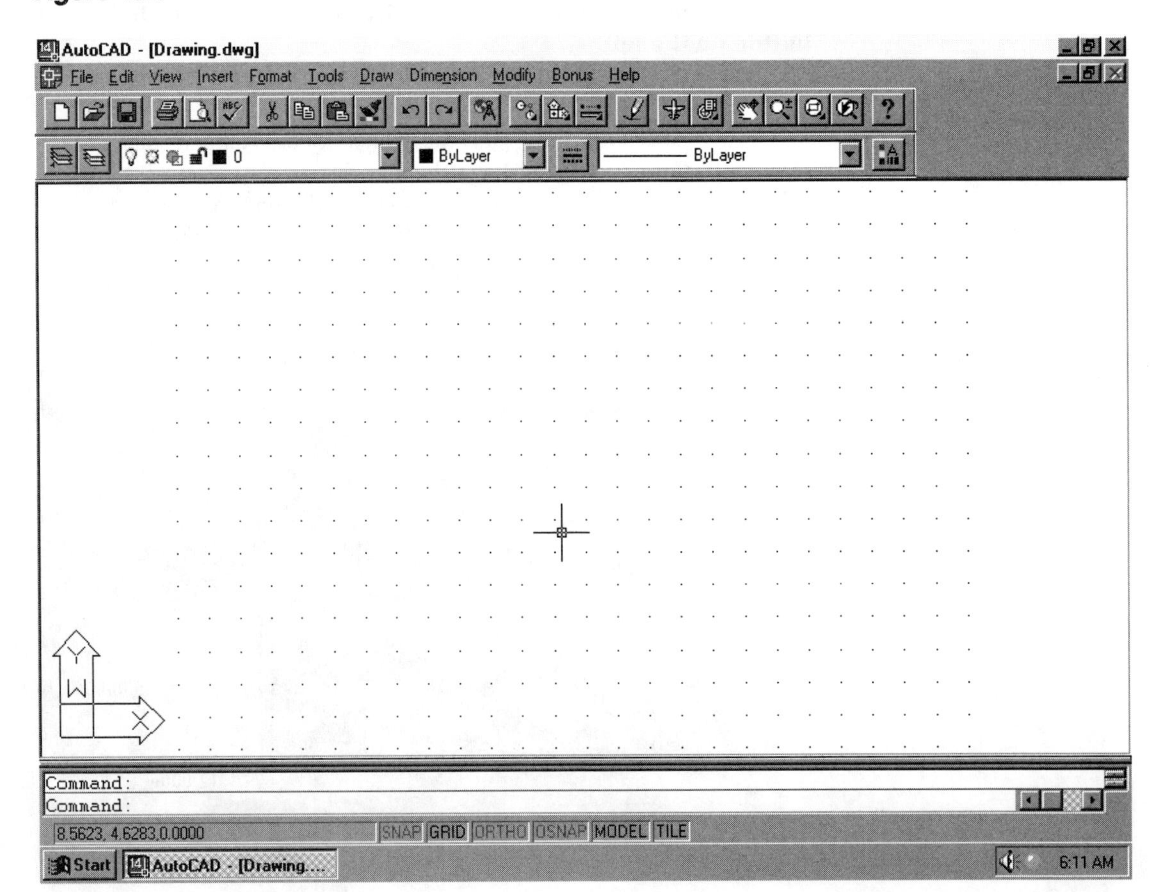

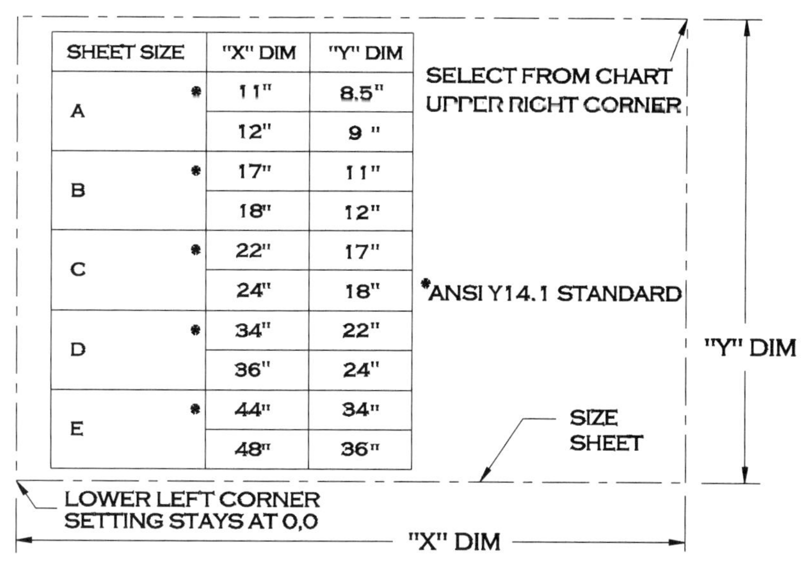

SHEET SIZE		"X" DIM	"Y" DIM
A	*	11"	8.5"
		12"	9 "
B	*	17"	11"
		18"	12"
C	*	22"	17"
		24"	18"
D	*	34"	22"
		36"	24"
E	*	44"	34"
		48"	36"

Figure 4.3.

values for the lower left and upper right corners of the grid in place of widths and lengths of the grid boundaries. To use LIMITS, follow this procedure:

1. Type "Limits" or select "Drawing Limits" from the Format menu.
2. Type coordinates for the lower left corner, for example (0,0).
3. Type coordinates for the upper right corner, for example (18,12).

The Off and On options in the LIMITS command control what happens when you try to draw outside of the defined limits. With LIMITS off, nothing will happen. With LIMITS on, AutoCAD will not accept any attempt to begin an entity outside of limits, notifying you with the message "Attempt to draw outside of limits". You can extend objects beyond the limits as long as they were started within them. By default, LIMITS is off.

⌗ When you are done experimenting, return LIMITS to (0,0) and (18,12), using the LIMITS command, and then ZOOM All.

You are now in the drawing that we will use for your template, so it is not necessary to begin a new drawing for the next section.

4.2 CREATING A TEMPLATE

To make your own template, so that new drawings will begin with the settings you want, all you have to do is create a drawing that has those settings and then tell AutoCAD that this is the drawing you want to use to define your initial drawing setup. The first part should be easy for you now, since you have been doing your own drawing setup for each new drawing in this book.

⌗ Make changes to the present drawing as follows:

GRID	1.00 ON (F7)	COORD	ON (F6)
SNAP	.25 ON (F9)	LTSCALE	.5
UNITS	2-place decimal	LIMITS	(0,0) (18,12)

NOTE: If you use the Quick Setup Wizard to set limits, you will still have to set the grid and snap as shown, since the wizard uses a .70 grid and snap spacing when you set to 18 × 12.

⌗ Load all linetypes from the ACAD file using the following procedure.

1. Open the Linetype dialog box from the Format menu.
2. Click on Load to open the Load dialog box.
3. Highlight the first Linetype on the list.
4. Scroll down to the end of the list.
5. Hold down the shift key as you highlight the last linetype on the list.
6. With all linetypes highlighted, click OK.
7. Click OK in the Layer and Linetype dialog box, or click the Layer tab to define layers.

⌗ Create the following layers and associated colors and linetypes.

Remember that you can make changes to your template at any time. The layers called "text", "hatch", and "dim" will not be used until Chapters 7 and 8, in which we introduce text, hatch patterns, and dimensions to your drawings. Creating them now will save time and avoid confusion later on.

Layer 0 is already defined.

Layer Name	*On/Off*	*Color*	*Linetype*
0	On	7(white)	Continuous
1	On	1(red)	Continuous
2	On	2(yellow)	Hidden
3	On	3(green)	Center
Text	On	4(cyan)	Continuous
Hatch	On	5(blue)	Continuous
Dim	On	6(magenta)	Continuous

⌗ When all changes are made, save your drawing as "B".

NOTE: Do not leave anything drawn on your screen or it will come up as part of the template each time you open a new drawing. For some applications this may be useful. For now we want a blank template.

If you have followed instructions up to this point, B.dwg should be on file. Now we will use it as the template for a new drawing.

4.3 SELECTING A TEMPLATE DRAWING

The procedure for designating a template drawing involves saving the drawing as a template file. The template file will be given a .dwt extension and placed in the template file folder. It will automatically become the default template until you select another template.

⌗ To begin this task you should be in the B drawing created in the last task. All the drawing changes should be made as described.

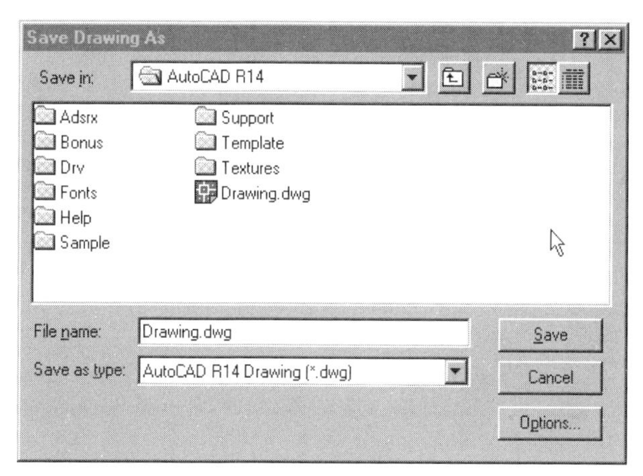

Figure 4.4.

⌖ **Open the File menu and select Save As. . . .**

This will call up the familiar Save Drawing As dialog box shown in *Figure 4.4.* The edit box in the middle holds the name of the current drawing. If you have not named the drawing, it will be called "Drawing".

⌖ **Double click in the File name box to highlight the drawing name.**

⌖ **Type "B" for the new name.**

Below the file name box is the Save as type box. This box will list four different options for saving the drawing. Drawing Template File is the last.

⌖ **Highlight "Drawing Template File (*.dwt)", and click on it so that it shows up in the list box.**

This will also open the Template file folder automatically. You will see that there are many templates already there. These are supplied by AutoCAD and may be useful to you later. At this point it is more important to learn how to make your own.

Once your drawing name ("B") is in the name box and the Save as type box shows Drawing Template File, you are ready to save.

⌖ **Click on Save.**

This will call up a Template Description box as illustrated in *Figure 4.5.* Here you can specify a description that will appear in the Create New Drawing dialog box whenever this template is used.

⌖ **Type "B size template, blank page".**

Figure 4.5.

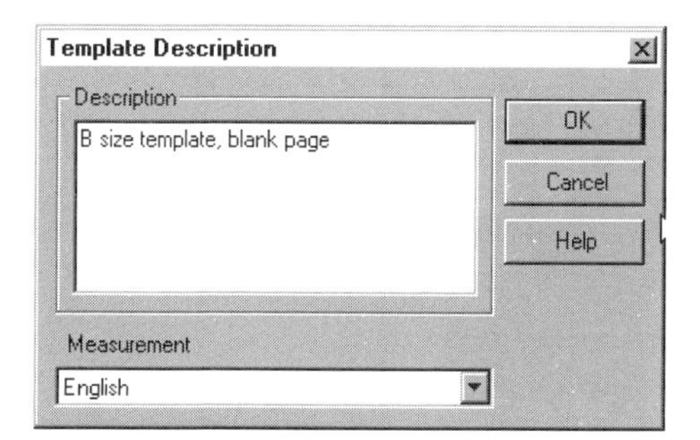

This will indicate that this is your B template and that there is no title block or other objects drawn in the template.

⌖ Click on OK.

The task of creating the drawing template is now complete. All that remains is to create a new drawing using the template to see how it works.

4.4 USING THE MOVE COMMAND

The ability to copy and move objects on the screen is one of the great advantages of working on a CAD system. It can be said that CAD is to drafting as word processing is to typing. Nowhere is this analogy more appropriate than in the "cut and paste" capacities that the COPY and MOVE commands give you.

⌖ Draw a circle with a radius of 1 somewhere near the center of the screen (9,6), as shown in *Figure 4.6.*

As discussed in Chapter 2, AutoCAD allows you to pick objects before or after entering an edit command. In this exercise we will use MOVE both ways, beginning with the verb/noun method. We will use the circle you have just drawn, but be aware that the selection set could include as many entities as you like, and that a group of entities can be selected with a window or crossing box.

⌖ Type "m", select Move from the Modify menu, or select the Move tool from the Modify toolbar, as shown in *Figure 4.7.*

You will be prompted to select objects to move.

⌖ Point to the circle.

As in the ERASE command, your circle will become dotted.

In the command area, AutoCAD will tell you how many objects have been selected and prompt you to select more. When you are through selecting objects you will need to press Enter to move on.

⌖ Press Enter.

AutoCAD prompts:

Base point or displacement:

Most often you will show the movement by defining a vector that gives the distance and direction you want the object to be moved. In order to define movement

Figure 4.6.

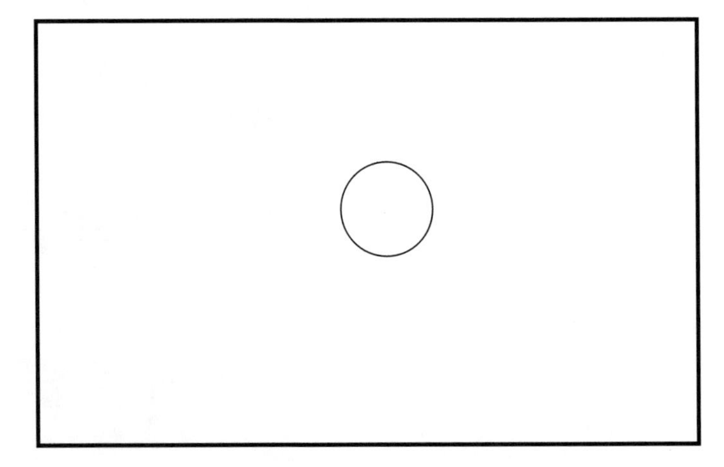

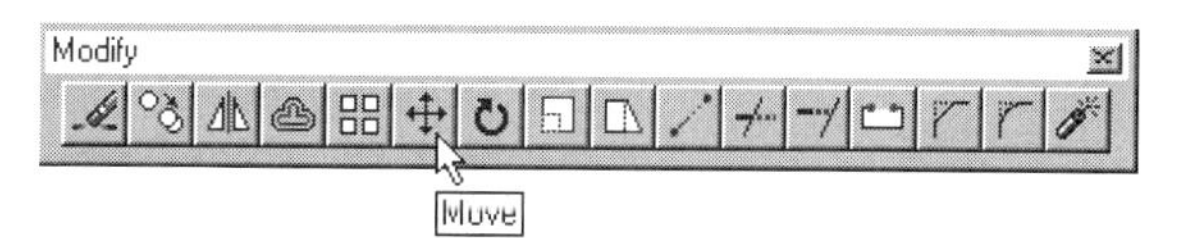

Figure 4.7.

with a vector, all AutoCAD needs is a distance and a direction. Therefore, the base point does not have to be on or near the object you are moving. Any point will do, as long as you can use it to show how you want your objects moved. This may seem strange at first, but it will soon become quite natural. Of course, you may choose a point on the object if you wish. With a circle, the center point may be convenient.

⌖ **Point to any location not too near the right edge of the screen.**

AutoCAD will give you a rubber band from the point you have indicated and will ask for a second point:

Second point of displacement:

As soon as you begin to move the cursor, you will see that AutoCAD also gives you a circle to drag so you can immediately see the effect of the movement you are indicating. An example of how this may look is shown in *Figure 4.8.* Let's say you want to move the circle 3.00 to the right. Watch the coordinate display and stretch the rubber band out until the display reads "3.00 < 0" (press F6 or double click on the coordinate display to get polar coordinates).

⌖ **Pick a point 3.00 to the right of your base point.**

The rubber band and your original circle disappear, leaving you a circle in the new location.

Now, if ortho is on, turn it off (F8) and try a diagonal move. This time we will use the previous option to select the circle.

⌖ **Type "m" or select the Move tool, or press Enter to repeat the command.**

AutoCAD follows with the "Select objects:" prompt.

⌖ **Reselect the circle by typing "p" for previous.**

⌖ **Press Enter to end the object selection process.**

⌖ **Select a base point.**

⌖ **Move the circle diagonally in any direction you like.**

Figure 4.8.

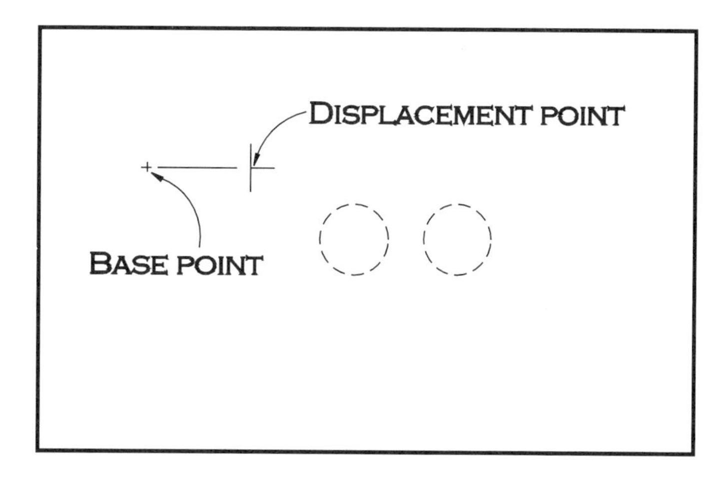

Figure 4.9 is an example of how this might look.

⌖ **Try moving the circle back to the center of the screen.**

It may help to choose the center point of the circle as a base point this time, and choose a point at or near the center of the grid for your second point.

Moving with Grips

You can use grips to perform numerous editing procedures without ever entering a command. This is probably the simplest of all editing methods, called "autoediting". It does have some limitations, however. In particular, you can only select by pointing, windowing, or crossing.

⌖ **Point to the circle.**

It will become highlighted and grips will appear.

Notice that grips for a circle are placed at quadrants and at the center. In more involved editing procedures the choice of which grip or grips to use for editing is significant. In this exercise you will do fine with any of the grips.

⌖ **Move the pickbox slowly over one of the grips.**

If you do this carefully you will notice that the pickbox "locks onto" the grip as it moves over it. You will see this more clearly if snap is off (F9).

⌖ **When the pickbox is locked on a grip, press the pick button.**

The selected grip will become filled and change colors (from blue to red).

In the command area you will see this:

```
** STRETCH **
<Stretch to point>/Base point/Copy/Undo/eXit:
```

Stretching is the first of a series of five autoediting modes that you can activate by selecting grips on objects. The word "stretch" has many meanings in AutoCAD and they are not always what you would expect. We will explore the stretch autoediting mode and the STRETCH command in Chapter 6. For now, we will bypass stretch and use the MOVE mode.

Release 14 has a convenient pop up menu for use in grip editing.

⌖ **Click the enter button on your mouse.**

This will call up the pop up menu shown in *Figure 4.10.* It contains all of the grip edit modes plus several other options.

Figure 4.9.

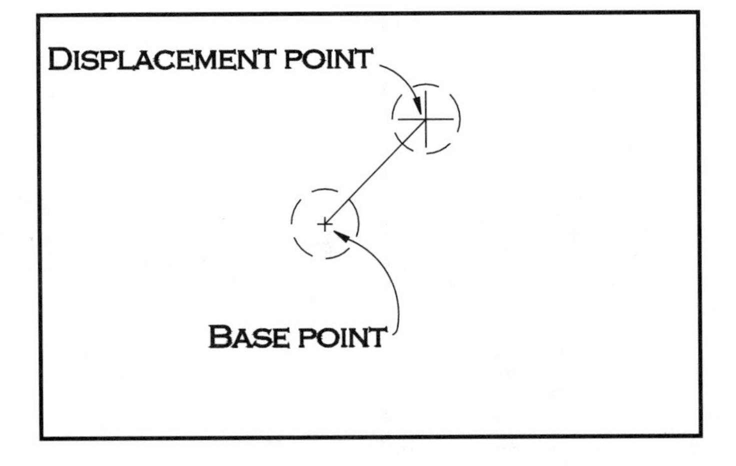

Figure 4.10.

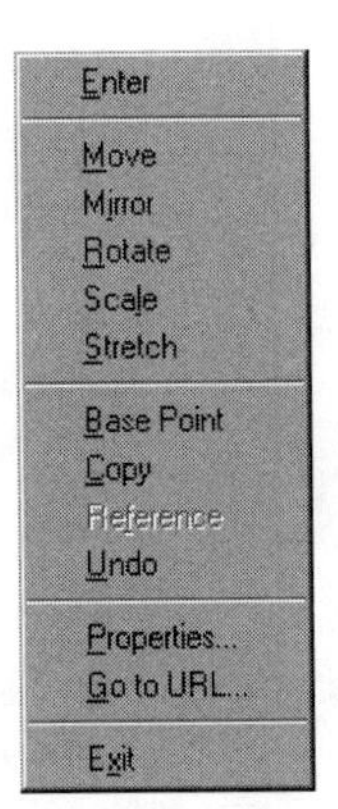

⌗ **Select Move in the pop up menu box.**

The pop up list will disappear and you will be in Move mode. Move the cursor and you will see a rubber band from the selected grip to the same position on a dragged circle.

⌗ **Pick a point anywhere on the screen.**

The circle will move where you have pointed.

4.5 USING THE COPY COMMAND

The COPY command works so much like the MOVE command that you should find it quite easy to learn at this point, and we will give you fewer instructions. The main difference is that the original object will not disappear when the second point of the displacement vector is given. Also, there is the additional option of making multiple copies of the same object, which we will explore in a moment.

But first, we suggest that you try making several copies of the circle in various positions on the screen. Try using both noun/verb and verb/noun sequences.

⌗ **To initiate COPY, type "co" or select Copy from the Modify menu, or the Copy tool from the Modify toolbar, as shown in *Figure 4.11*.**

Notice that "c" is not an alias for COPY. Also, notice that there is a Copy tp Clipboard tool on the Standard toolbar that initiates the COPYCLIP command. This tool has a very different function. It is used to copy objects to the Windows 95 Clipboard and then into other applications. It has no effect on objects within your drawing, other than to save them on the clipboard.

When you are satisfied that you know how to use the basic COPY command, move on to the Multiple copy option.

The Multiple Copy Option

This option allows you to show a series of vectors starting at the same base point, with AutoCAD placing copies of your selection set accordingly.

⌗ **Type "co" or select the Copy tool from the Modify toolbar.**
⌗ **Point to one of the circles on your screen.**
⌗ **Press Enter to end the selection process.**
⌗ **Type "m" for the Multiple option.**
⌗ **Show AutoCAD a base point.**
⌗ **Show AutoCAD a second point.**

You will see a new copy of the circle, and notice also that the prompt for a "Second point of displacement" has returned in the command area. AutoCAD is waiting for another vector, using the same base point as before.

⌗ **Show AutoCAD another second point.**
⌗ **Show AutoCAD another second point.**

Figure 4.11.

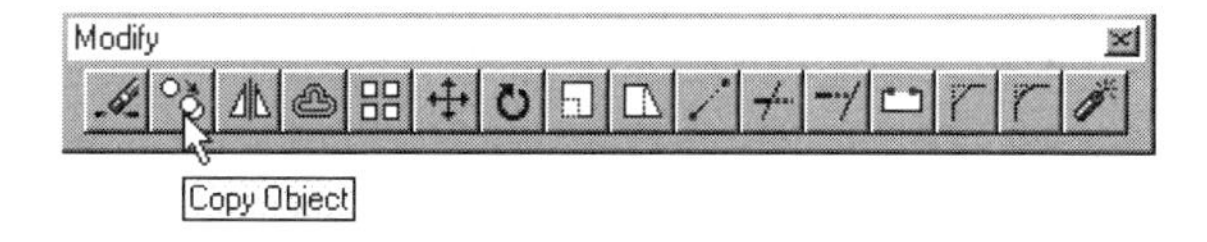

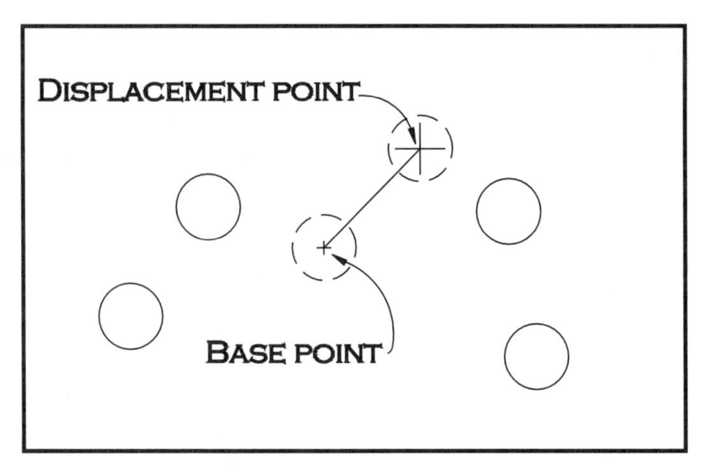

Figure 4.12.

Repeat as many times as you wish. If you get into this you may begin to feel like a magician pulling rings out of thin air and scattering them across the screen. The results will appear something like *Figure 4.12.*

Copying with Grips

The grip editing system includes a variety of special techniques for creating multiple copies in all five modes. The function of the Copy option will differ depending on the grip edit mode. For now we will use with the Copy option with the Move mode, which provides a shortcut for the same kind of process you just went through with the COPY command.

Since you should have several circles on your screen now, we will take the opportunity to demonstrate how you can use grips on more than one object at a time. This can be be very useful if your drawing contains two or more objects that maintain the same relationship with each other at different locations in your drawing.

Pick any two circles.

The circles you pick should become highlighted and grips should appear on both, as illustrated in *Figure 4.13.*

Figure 4.13.

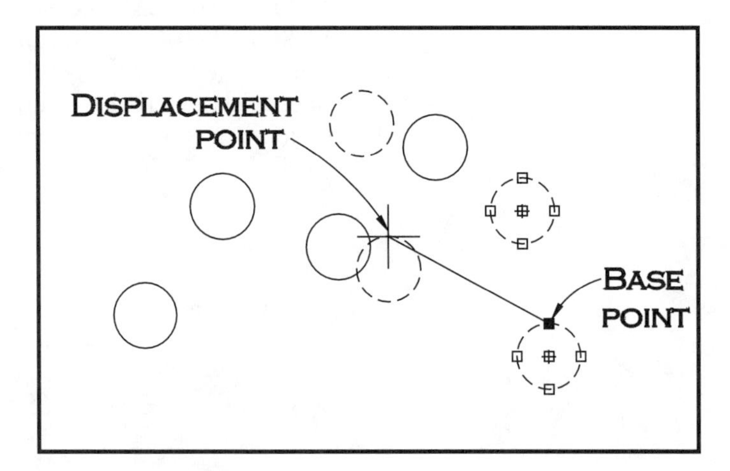

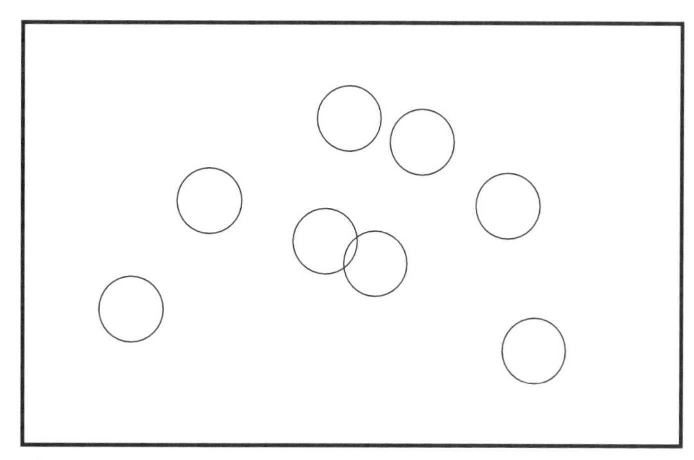

Figure 4.14.

☩ **Pick any grip on either of the two highlighted circles.**

The grip should change colors. This time we will not use the pop up menu. In the command area you will see the grip edit prompt for the Stretch mode:

** STRETCH **
<Stretch to point>/Base point/Copy/Undo/eXit:

☩ **Press the space bar enter button on your cursor.** (Do not press the enter key on your mouse.)

This will bring you to the MOVE mode prompt:

** MOVE**
<Move to point>/Base point/Copy/Undo/eXit:

☩ **Type "c" to initiate copying.**

The prompt will change to:

** MOVE (multiple) **
<Move to point>/Base point/Copy/Undo/eXit:

You will find that all copying in the grip editing system is multiple copying. Once in this mode AutoCAD will continue to create copies wherever you press the pick button until you exit by typing "x" or pressing Enter.

☩ **Move the cursor and observe the two dragged circles.**

☩ **Pick a point to create copies of the two highlighted circles, as illustrated in *Figure 4.14.***

☩ **Pick another point to create two more copies.**

☩ **When you are through, press Enter to exit the grip editing system.**

☩ **Press Esc twice to remove grips.**

4.6 USING THE ARRAY COMMAND—RECTANGULAR ARRAYS

The ARRAY command gives you a powerful alternative to simple copying. An array is a repetition in matrix form of the same figure. This command takes an object or group of objects and copies it a specific number of times in mathematically defined, evenly spaced, locations.

There are two types of arrays. Rectangular arrays are linear and defined by rows and columns. Polar arrays are angular and based on the repetition of objects around the circumference of an arc or circle. The dots on the grid are an example of a rectangular array; the lines on any circular dial are an example of a polar array. Both types are common. We will explore rectangular arrays in this chapter and polar arrays in the next.

In preparation for this exercise, erase all the circles from your screen. This is a good opportunity to try the ERASE All option.

- **Type "e" or select the Erase tool from the Modify toolbar.**
- **Type "all".**
- **Press Enter.**
- **Now draw a single circle, radius .5, centered at the point (2,2).**
- **Type "Ar" or select Array from the Modify menu, or the Array tool from the Modify toolbar, as shown in *Figure 4.15.***

You will see the "Select objects" prompt.

- **Point to the circle.**
- **Press Enter to end the selection process.**

AutoCAD will ask which type of array you want:

Rectangular or Polar array (R/P) <R>:

- **Type "r" or you can press Enter if R is the default.**

AutoCAD will prompt you for the number of rows in the array.

Number of rows (---) <1>:

The (---) is to remind you of what a row looks like, i.e., it is horizontal. The default is 1. So if you press Enter you will get a single row of circles. The number of circles in the row will depend, then, on the number of columns you specify. We will ask for three rows instead of just one.

- **Type "3".**

AutoCAD now asks for the number of columns in the array:

Number of columns (|||) <1>:

Using the same format (|||), AutoCAD reminds you that columns are vertical. The default is 1 again. What would an array with three rows and only one column look like?

We will construct a five-column array.

- **Type "5".**

Now AutoCAD needs to know how far apart to place all these circles. There will be 15 of them in this example—three rows with five circles in each row. AutoCAD prompts:

Unit cell or distance between rows (---):

"Unit cell" means that you can respond by showing two corners of a window. The horizontal side of this window would give the space between columns; the vertical

Figure 4.15.

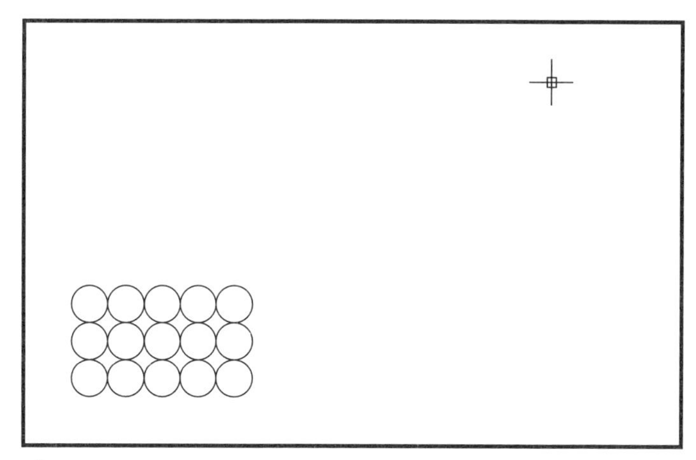

Figure 4.16.

side would give the space between rows. You could do this exercise by showing a 1 × 1 window. We will use the more basic method of typing values for these distances. The distance between rows will be a vertical measure.

⊕ **Type "1".**

AutoCAD now asks for the horizontal distance between columns:

Distance between columns (|||):

⊕ **Type "1" again.**

You should have a 3 × 5 array of circles, as shown in *Figure 4.16.*

Notice that AutoCAD builds arrays up and to the right. This is consistent with the coordinate system, which puts positive values to the right on the horizontal x axis, and upwards on the vertical y axis. Negative values can be used to create arrays in other directions.

4.7 CHANGING PLOT CONFIGURATION PARAMETERS

In the previous chapter you began using the plot preview feature of the Plot Configuration Dialog box. In this chapter we encourage you to continue to learn about plot configuration by exploring several more areas of the dialog box. We have used no specific drawing for illustration. Now that you know how to use plot preview, you can observe the effects of changing plot parameters with any drawing you like and decide at any point whether you actually want to print out the results. We will continue to remind you to look at a plot preview after making any change in configuration. This is the best way to learn about plotting.

We will start with device selection and move on to pen parameters and additional parameters. In the next chapter we will complete the introduction to the dialog box with paper size, orientation, scaling, rotation, and origin.

⊕ **To begin this exploration, you should have a drawing or drawn objects on your screen so that you can observe the effects of various configuration changes you will be making.**

⊕ **Type "Ctrl-p", select "Print" from the File pull down menu, or select the Print tool from the Standard toolbar.**

This will open up the Plot Configuration dialog box.

Device and Default Information

The device box should contain the name of the printer or plotter you are planning to use. If not, you can get a list of available options through the subdialog illustrated in *Figure 4.17.*

Click on "Device and Default Selection" or type "d".

You should see a dialog box similar to the one shown in the figure. The list of devices will be your own, of course. If there is a need to change plotters or printers you can do so by selecting from the list. If the device you want is not on the list, you will need to make sure the device driver is properly installed and then use the CONFIG command to add it to the list. See the *AutoCAD Installation Guide* for additional information.

Along with the list of available plotting devices, this dialog box allows you to save plotting specifications to ASCII files and get them back later. AutoCAD automatically saves your most recent plot parameters in a file called ACAD.cfg. However, for more permanent storage, you can create a separate file. Such files are saved with a "pc2" extension for complete configuration files or "pcp" (plot configuration parameter) for partial configuration files. There are many parameters to deal with in plotting, including as many as 256 pen colors. So saving a set of parameters to a file may be far more efficient than specifying them manually each time you plot. You may have one set of parameters for one device, and another set for a different device. Pcp files save only device independent information, while pc2 files save all configuration information.

In addition, some plotting devices have special configuration requirements not needed or not available on other devices. If this is the case with your device, the boxes under "Device Specific Configuration" will be accessible. The specific parameters and options will be determined by the device driver for your plotter. See the *AutoCAD Installation Guide* if you need additional information.

Click on OK or Cancel to exit the Device and Default Selection dialog.

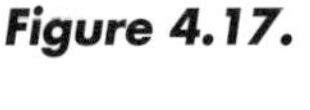

Figure 4.17.

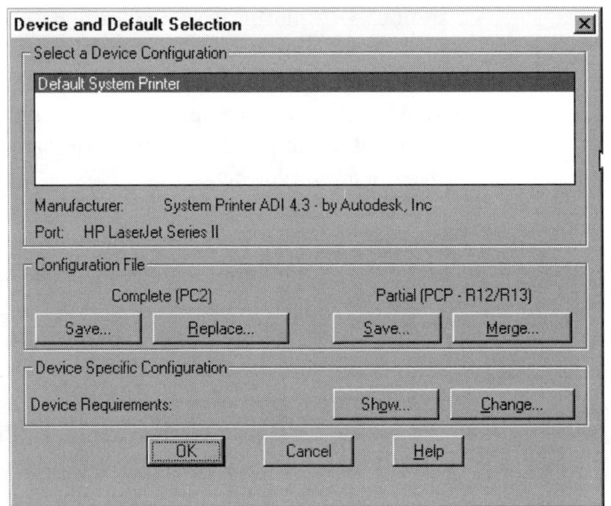

This will bring you back to the Plot Configuration dialog box. Assuming the device named in the Device and Default Information box is the correct one, you are ready to move on. However, if you have made any changes, we recommend that you do a plot preview. Changing devices may bring about changes in paper size and orientation, for example, and you will see these changes reflected graphically in the partial and full previews.

Pen Parameters

If you are using a plotter with multiple pens, you can assign colors, linetypes, widths, and speeds to each pen individually. If you are using a printer, you will find that the Pen Assignments box is grayed out. If you do have access, we suggest you look at the Pen Assignment dialog now, even though you may not want to make changes.

⌖ **Click on "Pen Assignments" or type "p".**

This will call the dialog box shown in *Figure 4.18.* If you select any pen, you will see its color, number, linetype, speed, and width displayed in the Modify Values box at the right. Changes can be made in these edit boxes in the usual manner.

The specifications made here are designed to relate pen numbers on your plotter to pen colors. If pen numbers are correctly related to pen colors, then layers will automatically be plotted in their assigned colors. The linetypes assigned to layers in AutoCAD are also plotted automatically and do not need to be assigned to pens at this point.

NOTE: If your plotter supports multiple linetypes, you will reach another dialog box showing numbered linetypes by clicking on "Feature Legend . . .?". These are not to be confused with the linetypes created within your AutoCAD drawing associated with layers. They should be used in special applications to vary the look of "continuous" AutoCAD lines only. Otherwise you will get a confused mixture of linetypes when your plotter tries to break up lines that AutoCAD has already drawn broken.

⌖ **Click on OK or Cancel to exit the "Pen Assignments" dialog box.**

Additional Parameters

This is a crucial area of the dialog box that allows you to specify the portion of your drawing to be plotted. You have some familiarity with this from Chapter 2, where you

Figure 4.18.

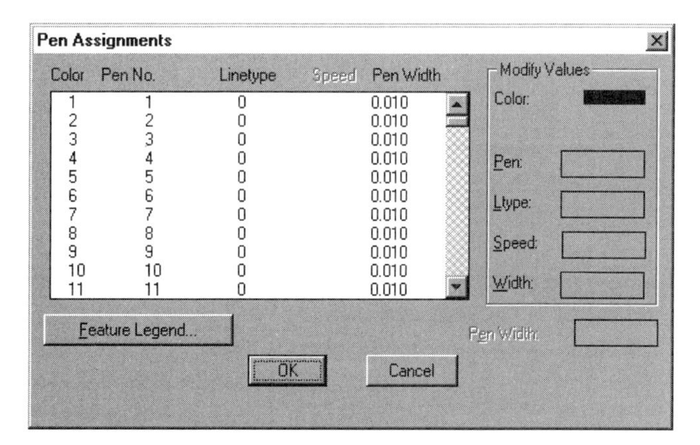

plotted using a window selection. Changes here will have a very significant impact on the effective plotting area. Be sure to use plot preview any time you make changes in these parameters.

Also available in this area are hidden line removal (for 3D drawings), fill area adjustment, which affects the way pen width is interpreted when drawing wide lines, and plotting to a file instead of to an actual plotter. These are discussed briefly at the end of this section.

The radio buttons on the left show the options for plotting area. "Display" will create a plot using whatever is actually on the screen. If you used the ZOOM command to enlarge a portion of the drawing before entering PLOT and then selected this option, AutoCAD would plot whatever you have zoomed in on. "Extents" refers to the actual drawing area in which you have drawn objects. It may be larger or smaller than the limits of the drawing. Drawing limits, as you know, are specified using the LIMITS command or a setup Wizard. If you are using our standard B-size template, they will go from (0,0) to (18,12). "View" will be accessible only after you have defined views in the drawing using the VIEW command (Chapter 11). "Window" will not be accessible until you define a window (if you have defined a window in a previous drawing, it may be saved in ACAD.cfg, the plot configuration file).

As an exercise, we suggest that you try switching among Display, Extents, Limits, and Window selections and use plot preview to see the results. Use both partial and full previews to ensure that you can clearly see what is happening.

Whenever you make a change, also observe the changes in the boxes showing Plotted Inches = Drawing Units. Assuming that "Scaled to Fit" is checked, you will see significant changes in these scale ratios as AutoCAD adjusts scales according to how much area it is being asked to plot.

We will explore the last two areas of Plot Configuration, "Paper Size and Orientation" and "Scale, Rotation, and Origin" in the next chapter. In the meantime, you should continue to experiment with plotting and plot previewing as you complete the drawings at the end of this chapter.

4.8 REVIEW MATERIAL

Questions

1. What command is automated through the step two (area) tab of the Quick Setup Wizard?
2. Name at least five settings that would typically be included in a Template drawing.
3. Where are template drawings stored in a standard AutoCAD file configuration? What extension is given to template file names?
4. What is the value of using a template drawing?
5. What is the main difference between the command procedure for MOVE and COPY?
6. What is the main limitation of grip editing?
7. What do you have to do to remove grips from an object once they are displayed?

8. How do you access the grip edit pop up menu?
9. Explain how arrays are a special form of copying.
10. What is a rectangular array? What is a polar array?
11. Why is it important to do a partial plot preview after changing printing devices or after switching radio buttons in the additional parameters area of the plot configuration dialog box?

COMMANDS

Modify	Data
ARRAY (rectangular)	LIMITS
COPY	
MOVE	

Drawing Problems

1. Create a C size drawing template using an ANSI standard sheet size, layers and other settings as shown in this chapter. Use your B template to make this process easier.
2. Open a drawing with your new C size template and draw a circle with a 2 unit radius centered at (11,8).
3. Using grips make four copies of the circle, centered at (15,8), (11,12), (7,8) and (11,4).
4. Switch to layer 2 and draw a 1 × 1 square with lower left corner at (1,1).
5. Create a rectangular array of the square with 14 rows and 20 columns, 1 unit between rows, and 1 unit between columns.

PROFESSIONAL SUCCESS

What Is the Correct Procedure?

A group of twenty students given an AutoCAD assignment would be expected to produce drawings that are nearly identical to each other. It would be surprising, however, if any of them used the exact same sequence of commands to produce their drawings.

It is very important to understand that AutoCAD is a means to an end - that is, it is the appearance and precision of the final drawing that matters, and not the specific sequence of commands used to produce it. For example, there are at least twelve possible command sequences that can be used to draw a rectangle:

Lines may be used to draw the four sides separately. The "line" command may be chosen from either the pull-down menus or the toolbox, or it may be typed directly into the command area.

The complete rectangle may be drawn at once by using any of the three above methods to invoke the "rectangle" command.

Coordinates necessary to complete either the "line" or "rectangle" commands may either be typed in on the command line, or specified directly via the mouse or tablet.

DRAWING 4—1

Grill

This drawing should go very quickly if you use the ARRAY command.

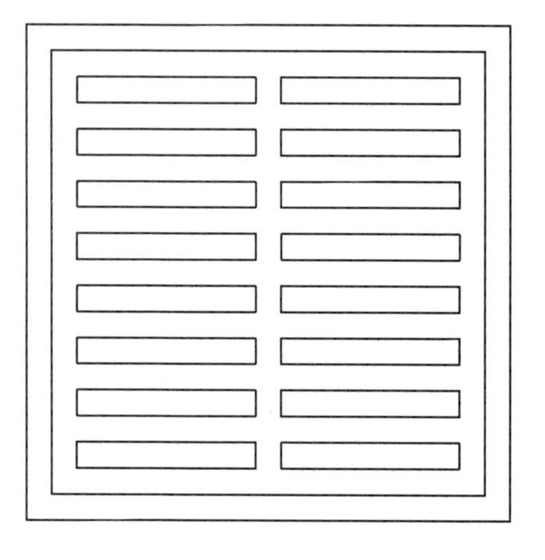

Drawing Suggestions

GRID = .5
SNAP = .25

- Begin with a 4.75 × 4.75 square.
- Move in .25 all around to create the inside square.
- Draw the rectangle in the lower left-hand corner first, then use the ARRAY command to create the rest.
- Also remember that you can undo a misplaced array using the U command.

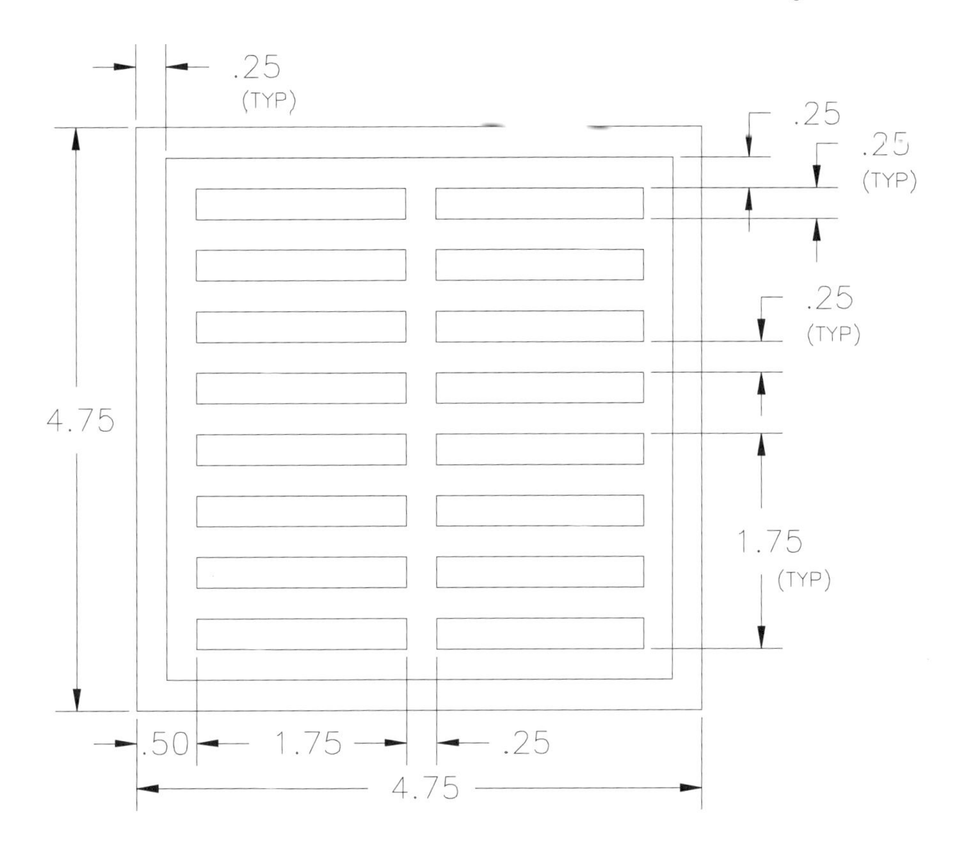

GRILL
Drawing 4–1

DRAWING 4—2

Test Bracket

This is a great drawing for practicing much of what you have learned up to this point. Notice the suggested snap, grid, ltscale, and limit settings, and use the ARRAY command to draw the 25 circles on the front view.

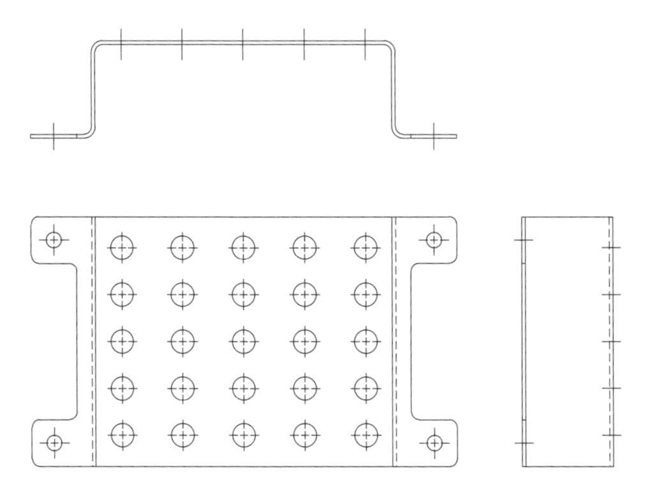

Drawing Suggestions

GRID = .5 LTSCALE = .50
SNAP = .25 LIMITS = (0,0) (24,18)

- Be careful to draw all lines on the correct layers, according to their linetypes.
- Draw center lines through circles before copying or arraying them, otherwise you will have to go back and draw them on each individual circle or repeat the array process.
- A multiple copy will work nicely for the four .50 diameter holes. A rectangular array is definitely desirable for the twenty-five .75 diameter holes.

Creating Center Marks with the Dimcen System Variable There is a simple way to create the center marks and center lines shown on all the circles in this drawing. It involves changing the value of a dimension variable called "dimcen" (dimension center). Dimensioning and dimension variables are discussed in Chapter 8, but if you would like to jump ahead, the following procedure will work nicely in this drawing.

1. Type "dimcen".
2. The default setting for dimcen is .09, which will cause AutoCAD to draw a simple cross as a center mark. Changing it to 2.09 will tell AutoCAD to draw a cross that reaches across the circle.
3. Type "2.09".
4. After drawing your first circle, and before arraying it, type "dim". This will put you in the dimension command.
5. Type "cen", indicating that you want to draw a center mark. This is a very simple dimension feature.
6. Point to the circle.

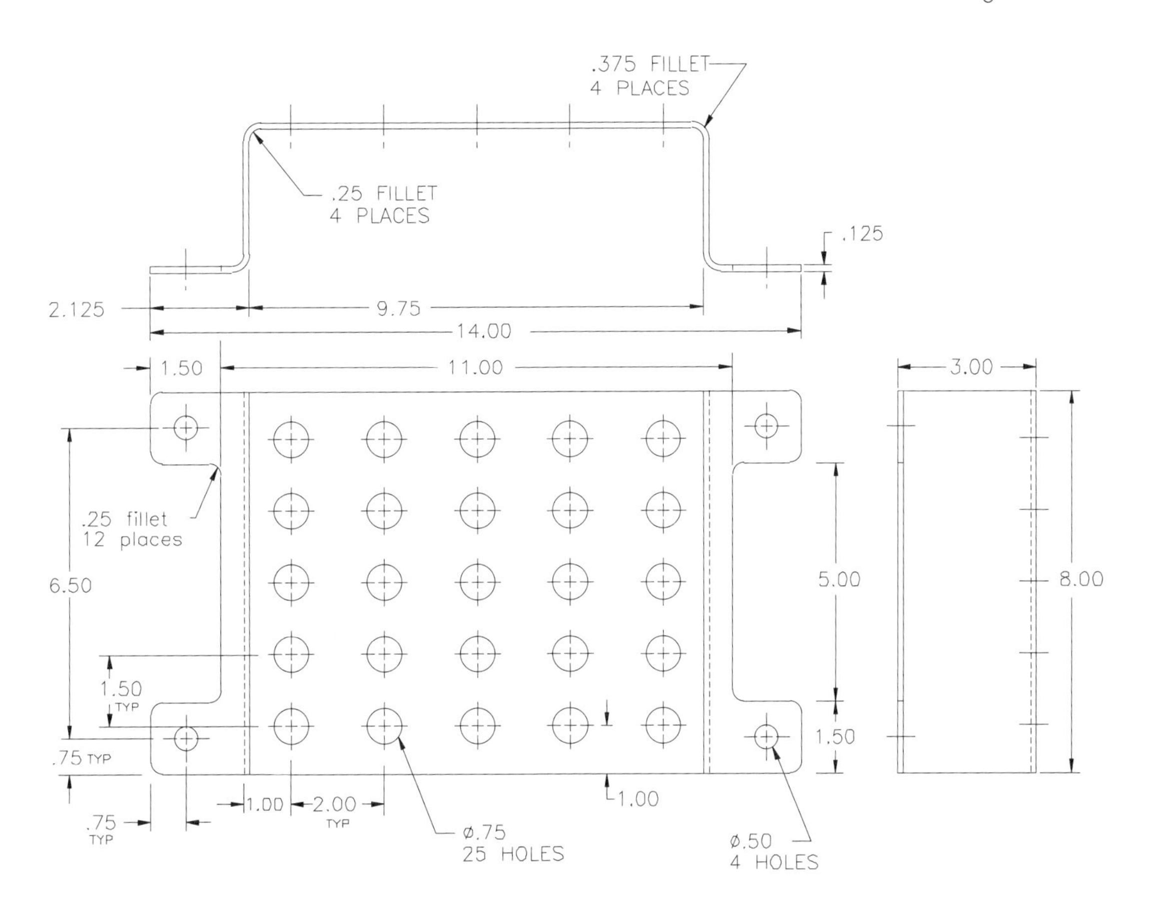

TEST BRACKET

Drawing 4–2

DRAWING 4—3

Floor Framing

This architectural drawing will require changes in many features of your drawing setup. Pay close attention to the suggested settings.

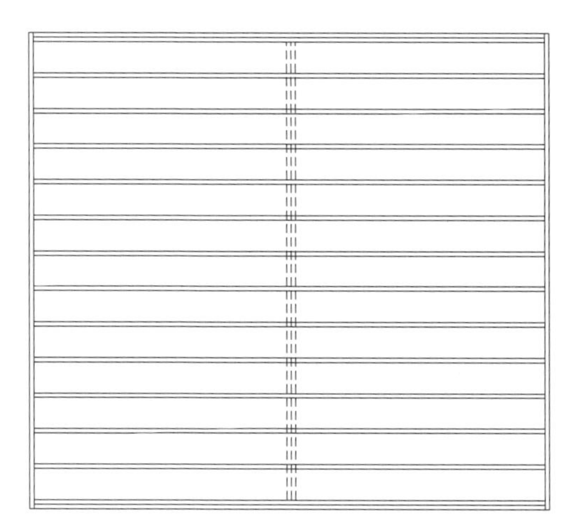

Drawing Suggestions

UNITS = Architectural, smallest fraction = 1"
LIMITS = 36", 24"
GRID = 1'
SNAP = 2"
LTSCALE = 12

- Be sure to use foot (') and inch (") symbols when setting limits, grid, and snap (but not ltscale).
- Begin by drawing the 20' × 17'-10" rectangle, with the lower left corner somewhere in the neighborhood of (4',4').
- Complete the left and right 2 × 10 joists by copying the vertical 17'-10" lines 2" in from each side. You may find it helpful to use the arrow keys when working with such small increments.
- Draw a 19'-8" horizontal line 2" up from the bottom and copy it 2" higher to complete the double joists.
- Array the inner 2 × 10 in a 14-row by 1-column array, with 16" between rows.
- Set to layer 2 and draw the three 17'-4" hidden lines down the center.

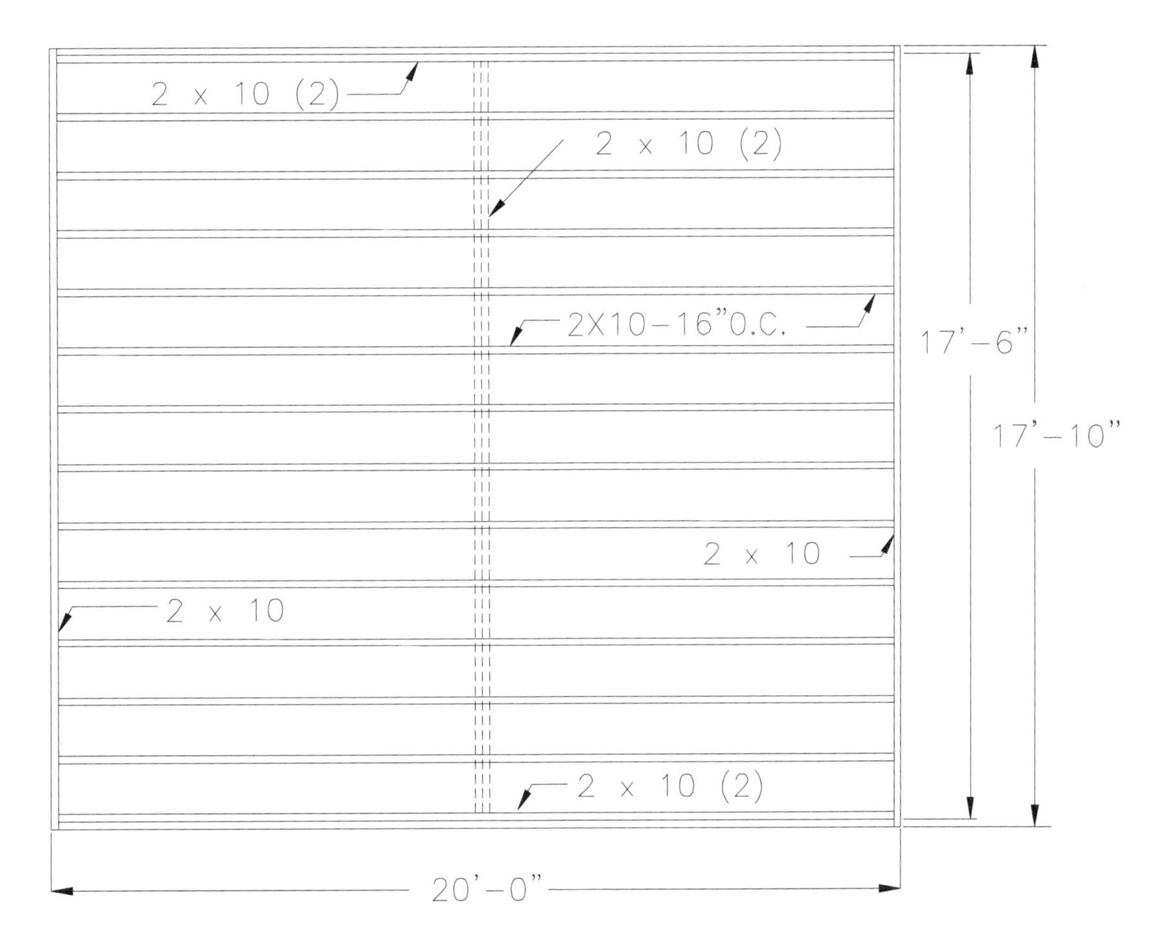

FLOOR FRAMING

Drawing 4–3

5

Arcs and Polar Arrays

Overview

So far, every drawing you have done has been composed of lines and circles. In this chapter you will learn a third major entity, the ARC. In addition, you will expand your ability to manipulate objects on the screen. You will learn to ROTATE objects and create their MIRROR images. But first, we will pick up where we left off in Chapter 4 by showing you how to create polar arrays.

Sections

Objectives

After reading this chapter, you should be able to:

- Create three polar arrays.
- Draw arcs in eight different ways.
- Rotate a previously drawn object.
- Create mirror images of previously drawn objects.
- Change Paper Size, Orientation, Plot Scale, Rotation, and Origin.
- Draw dials.
- Drawing an alignment wheel.
- Draw a hearth.

5.1 CREATING POLAR ARRAYS

The procedure for creating polar arrays is lengthy and requires some explanation. The first two steps are the same as in rectangular arrays. Step 3 is also the same, except that you respond with "p" instead of "r". From here on the steps will be new. First you will pick a center point, and then you will have several options for defining the array.

There are three qualities that define a polar array, but two are sufficient. A polar array is defined by two of the following: 1) a certain number of items, 2) an angle that these items span, and 3) an angle between each item and the next. However you define your polar array, you will have to tell AutoCAD whether or not to rotate the newly created objects as they are copied.

Begin a new drawing using the B template.

In preparation for this exercise, draw a vertical 1.00 line at the bottom center of the screen, near (9.00,2.00), as shown in *Figure 5.1.*

Figure 5.1.

We will use a 360 degree polar array to create *Figure 5.2.*

☩ **Type "Ar", select Array from the Modify menu, or the Array tool from the Modify toolbar**

☩ **Select the line.**

☩ **Type "p".**

So far, so good. Nothing new up to this point. Now you have a prompt that looks like this:

Base/Specify center point of array:

Rectangular arrays are not determined by a center, so we did not encounter this prompt before. Polar arrays, however, are built by copying objects around the circumferences of circles or arcs, so we need a center to define one of these. We will also need to specify whether objects should be rotated as they are copied. The Base option would allow us to determine how objects are rotated. For this array, show a center point.

☩ **Pick a point directly above the line and somewhat below the center of the screen.**

Something in the neighborhood of (9.00,4.50) will do. The next prompt is:

Figure 5.2.

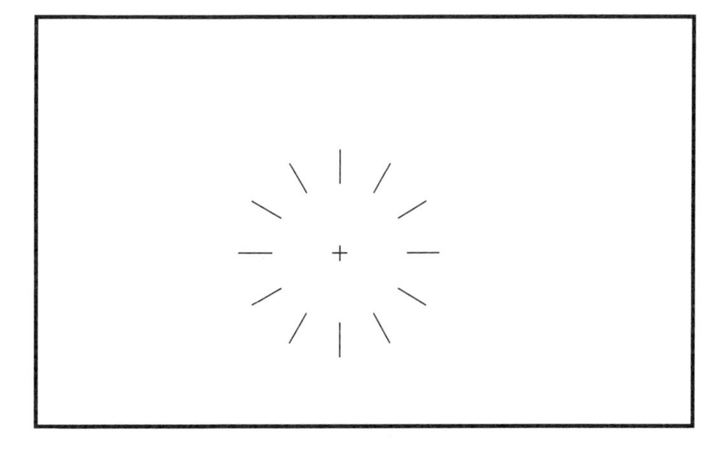

Number of items:

Remember that we have a choice of two out of three among number of items, angle to fill, and angle between items. This time we will give AutoCAD the first two.

⌗ Type "12".

Now that AutoCAD knows that we want 12 items, all it needs is either the angle to fill with these, or the angle between the items. It will ask first for the angle to fill:

Angle to fill (+=ccw,–=cw) <360>:

The symbols in parentheses tell us that if we give a positive angle the array will be constructed counterclockwise; if we give a negative angle, it will be constructed clockwise. Get used to this; it will come up frequently.

The default is 360 degrees, meaning an array that fills a complete circle.

If we did not give AutoCAD an angle (that is, if we responded with a "0"), we would be prompted for the angle between. This time around we will give 360 as the angle to fill.

⌗ Press Enter to accept the default, a complete circle.

AutoCAD now has everything it needs, except that it doesn't know whether we want our lines to retain their vertical orientation or to be rotated along with the angular displacement as they are copied. AutoCAD asks:

Rotate objects as they are copied? <Y>:

Notice the default, which we will accept.

⌗ Press Enter or type "y".

Your screen should resemble *Figure 5.2.*

This ends our discussion of polar arrays. With the options AutoCAD gives you there are many possibilities that you may want to try out. As always, we encourage experimentation. When you are satisfied, erase everything on the screen in preparation for learning the ARC command.

5.2 DRAWING ARCS

Learning AutoCAD's ARC command is an exercise in geometry. In this section we will give you a firm foundation for understanding and drawing arcs so that you will not be confused by all the options that are available. The information we give you will be more than enough to do the drawings in this chapter and most drawings you will encounter elsewhere. Refer to the *AutoCAD Command Reference* and the chart at the end of this section (*Figure 5.8*) if you need additional information.

AutoCAD gives you eight distinct ways to draw arcs, eleven if you count variations in order. Every option requires you to specify three pieces of information: where to begin the arc, where to end it, and what circle it is theoretically a part of.

We will begin by drawing an arc using the simplest method, which is also the default, the 3-points option. The geometric key to this method is that any three points not on the same line determine a circle or an arc of a circle. AutoCAD uses this in the CIRCLE command (the 3P option) as well as in the ARC command.

⌗ Type "a" or select the Arc tool from the Draw toolbar, as shown in *Figure 5.3*.

AutoCAD's response will be this prompt:

Center/<Start point>:

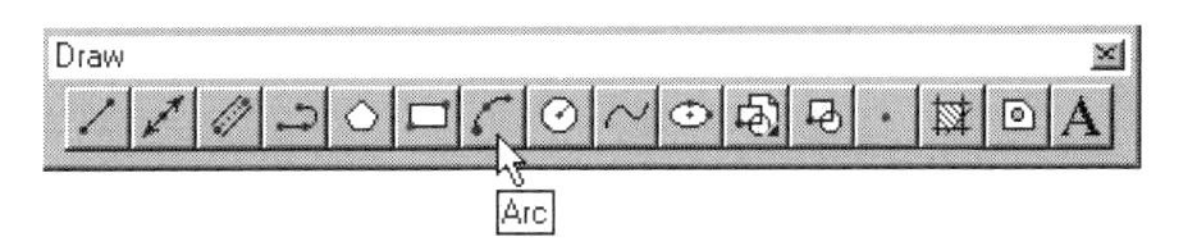

Figure 5.3.

Accepting the default by specifying a point will leave open all those options in which the start point is specified first.

If you instead type a "c", AutoCAD will prompt for a center point and follow with those options that begin with a center.

⌗ **Select a starting point near the center of the screen.**

AutoCAD prompts:

Center/End/<Second point>:

We will continue to follow the default three-point sequence by specifying a second point. You may want to refer to the chart (*Figure 5.4*) as you draw this arc.

⌗ **Select any point one or two units away from the previous point. Exact coordinates are not important.**

Once AutoCAD has two points, it gives you an arc to drag. By moving the cursor slowly in a circle and in and out you can see the range of what the third point will produce.

AutoCAD also knows now that you have to provide an end point to complete the arc, so the prompt has only one option:

End point:

Any point you select will do, as long as it produces an arc that fits on the screen.

⌗ **Pick an end point.**

As you can see, three-point arcs are easy to draw. It is much like drawing a line, except that you have to specify three points instead of two. In practice, however, you do not always have three points to use this way. This necessitates the broad range of options in the ARC command. The dimensions you are given and the objects already drawn will determine what options are useful to you.

Next, we will create an arc using the start, center, end method, the second option illustrated in *Figure 5.4*.

⌗ **Type "U" to undo the three-point arc.**

⌗ **Type "a" or select the Arc tool.**

⌗ **Select a point near the center of the screen as a start point.**

The prompt that follows is the same as for the three-point option, but we will not use the default this time:

Center/End/<Second point>

We will choose the Center option.

TIP: If you choose options from the Draw menu, you will find some steps automated. If you select "Start,Center,End" for example, the "c" will be entered automatically.

⌗ **Type "c", if necessary.**

TYPE	APPEARANCE	DESCRIPTION
3-point	2nd point; 1st point; 3rd point	Clockwise or counterclockwise
S,C,E (start, center, end)	start; end; center	Counterclockwise Radial rubber band indicates angle only, length is insignificant
S,C,A (start, center, angle)	start; 45°; center; 45°; ANGLE	+ angle = CCW − angle = CW Rubber band shows angle only, starting from horizontal
S,C,L (start, center, length of chord)	start; center; length of cord	Counterclockwise "Chord" rubber band shows length of chord only, direction is insignificant
S,E,A (start, end, angle)	90°; ANGLE; start; end	+ angle = CCW − angle = CW Rubber band shows angle only, starting from horizontal
S,E,R (start, end, radius)	radius = +2; start; end; radius = −2	Counterclockwise + radius = minor arc − radius = major arc Rubber band shows + radius values only, For − radius (type value)
S,E,D (start, end, direction)	end; direction; start	Direction of rubber band is a line tangent to the arc being constructed and runs through the start point
CONTIN: (continuous from line)	start; end	Arc begins at end point of previous line or arc and is tangent to it; Rubber band is a chord from start point to end point

Figure 5.4.

This tells AutoCAD that we want to specify a center point next, so we see this prompt:

Center:

⊕ **Select any point roughly one to three units away from the start point.**

The circle from which the arc is to be cut is now clearly determined. All that is left is to specify how much of the circle to take, which can be done in one of three ways, as the prompt indicates:

Angle/Length of chord/<End point>:

We will simply specify an end point by typing coordinates or pointing. But first, move the cursor slowly in a circle and in and out to see how the method works. As before, there is an arc to drag, and now there is a radial direction rubber band as well. If you pick a point anywhere along this rubber band, AutoCAD will assume you want the point where it crosses the circumference of the circle.

NOTE: Here, as in the polar arrays in this chapter, AutoCAD is building arcs counter-clockwise, consistent with its coordinate system.

⊕ **Select an end point to complete the arc.**

Now that you have tried three of the basic methods for constructing an arc, we strongly suggest that you study the chart and then try out the other options. The notes in the right-hand column will serve as a guide to what to look for.

The differences in the use of the rubber band from one option to the next can be confusing. You should understand, for instance, that in some cases the linear rubber band is only significant as a distance indicator; its angle is of no importance and is ignored by AutoCAD. In other cases it is just the reverse; the length of the rubber band is irrelevant, while its angle of rotation is important.

TIP: One additional trick you should try out as you experiment with arcs is as follows: If you press Enter or the space bar at the "Center/<Start point>" prompt, AutoCAD will use the end point of the last line or arc you drew as the new starting point and construct an arc tangent to it. This is the same as the Continue option on the pull down menu.

This completes the present discussion of the Arc command. Constructing arcs, as you may have realized, can be tricky. Another option that is available and often useful is to draw a complete circle and then use the TRIM or BREAK commands to cut out the arc you want. BREAK and TRIM are introduced in the next chapter.

5.3 USING THE ROTATE COMMAND

ROTATE is a fairly straightforward command, and it has some uses that might not be apparent immediately. For example, it frequently is easier to draw an object in a horizontal or vertical position and then ROTATE it rather than drawing it in a diagonal position.

In addition to the ROTATE command there is also a rotate mode in the grip edit system, which we will introduce at the end of the exercise.

⊕ **In preparation for this exercise, clear your screen and draw a horizontally oriented arc near the center of your screen, as in *Figure* 5.5. Exact coordinates and locations are not important.**

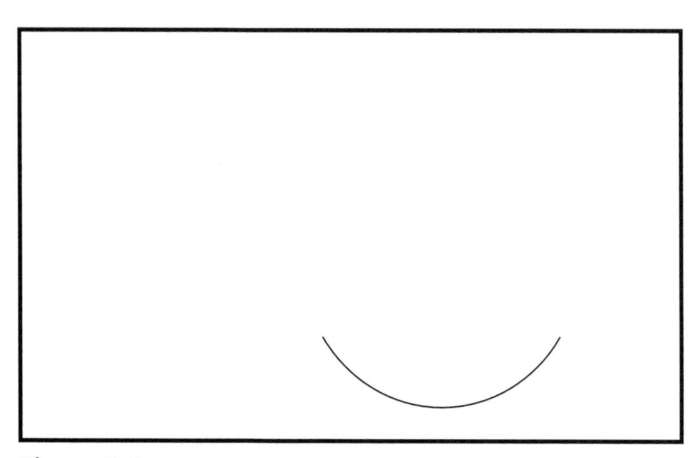

Figure 5.5.

We will begin by rotating the arc to the position shown in *Figure 5.6.*

⌖ **Select the arc.**

⌖ **Type "Ro", select Rotate from the Modify menu, or the Rotate tool from the Modify toolbar, as shown in *Figure 5.7.***

You will be prompted for a base point.

Base point:

This will be the point around which the object is rotated. The results of the rotation, therefore, are dramatically affected by your choice of base point. We will choose a point at the left tip of the arc.

⌖ **Point to the left tip of the arc.**

The prompt that follows looks like this:

<Rotation angle>/Reference:

The default method is to indicate a rotation angle directly. The object will be rotated through the angle specified and the original object deleted.

Figure 5.6.

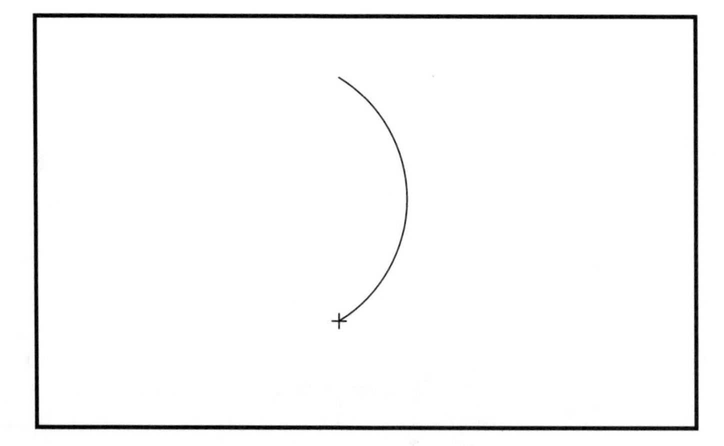

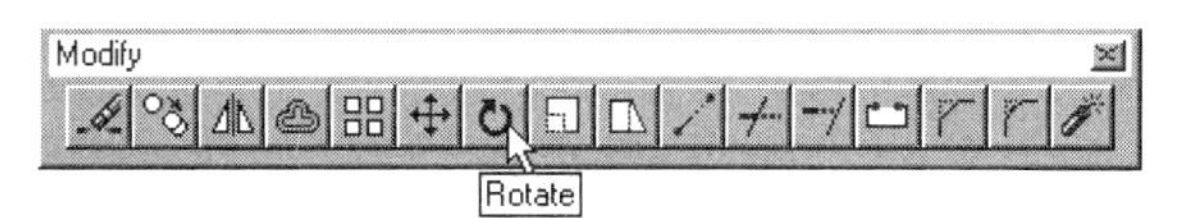

Figure 5.7.

Move the cursor in a circle and you will see that you have a copy of the object to drag into place visually. If ortho or snap are on, turn them off to see the complete range of rotation.

☩ Type "90" or point to a rotation of 90 degrees (use F6 if your coordinate display is not showing polar coordinates).

The results should resemble *Figure* 5.6.

Notice that when specifying the rotation angle directly like this, the original orientation of the selected object is taken to be 0 degrees. The rotation is figured counterclockwise from there. However, there may be times when you want to refer to the coordinate system in specifying rotation. This is the purpose of the "Reference" option. To use it, all you need to do is specify the present orientation of the object relative to the coordinate system, and then tell AutoCAD the orientation you want it to have after rotation. Look at *Figure* 5.8. To rotate the arc as shown, you either can indicate a rotation of –45 degrees or tell AutoCAD that it is presently oriented to 90 degrees and you want it rotated to 45 degrees. Try this method for practice.

☩ Press Enter to repeat the ROTATE command.
☩ Select the arc.
☩ Press Enter to end selection.
☩ Choose a base point at the lower tip of the arc.
☩ Type "r".

AutoCAD will prompt for a reference angle:

Reference angle <0>:

☩ Type "90".

AutoCAD prompts for an angle of rotation:

Figure 5.8.

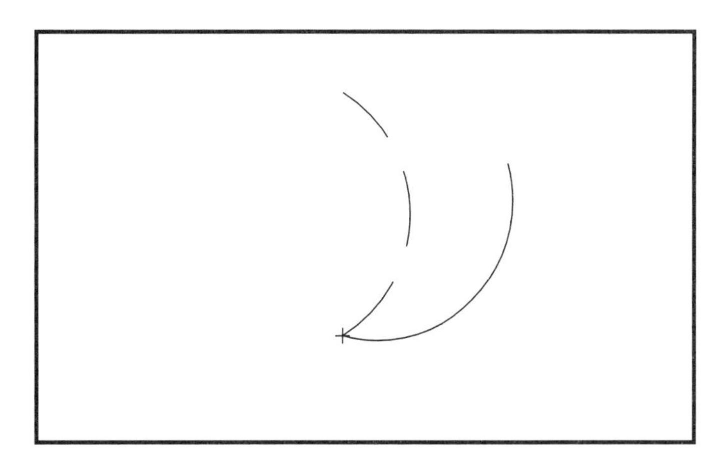

Rotation angle:

⌗ **Type "45".**

Your arc should now resemble the solid arc in *Figure* 5.8.

Rotating with Grips

Rotating with grips is simple and there is a very useful option for copying, but your choice of object selection methods is limited, as always, to pointing and windowing. Try this:

⌗ **Pick the arc.**

The arc will become highlighted and grips will appear.

⌗ **Pick the center grip.**

⌗ **Press the enter (right) button on your pointing device.**

⌗ **Select "Rotate" from the pop up menu.**

Move your cursor in a circle and you will see the arc rotating around the grip at the center of the arc.

⌗ **Now, type "b" or press the right button again and select "Base pt" from the pop up menu.**

Base point will allow you to pick a base point other than the selected grip.

⌗ **Pick a base point above and to the left of the grip, as shown in *Figure* 5.9.**

Move your cursor in circles again. You will see the arc rotating around the new base point.

⌗ **Type "c" or open the pop up menu and select "Copy".**

Notice the Command area prompt, which indicates you are now in a rotate and multiple copy mode.

⌗ **Pick a point showing a rotation angle of 90 degrees, as illustrated by the top arc in *Figure* 5.9.**

⌗ **Pick a second point showing a rotation angle of 180 degrees, as illustrated by the arc at the left in the figure.**

⌗ **Pick point 3 at 270 degrees to complete the design shown in *Figure* 5.9.**

Figure 5.9.

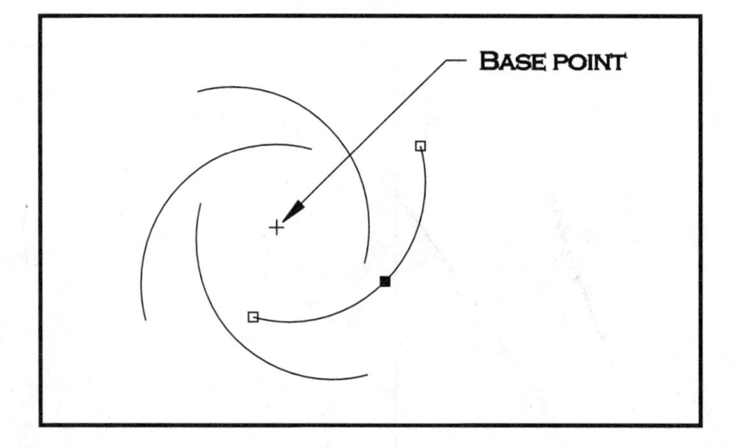

⌗ **Press enter or the space bar to exit the grip edit mode.**

This capacity to create rotated copies is very useful, as you will find when you do the drawings at the end of the chapter.

5.4 CREATING MIRROR IMAGES OF OBJECTS ON THE SCREEN

There are two main differences between the command procedures for Mirror and Rotate. First, in order to mirror an object you will have to define a mirror line, and second you will have an opportunity to indicate whether you want to retain the original object or delete it. In the Rotate sequence the original is always deleted.

There is also a mirror mode in the grip edit system, which we will explore at the end of the task.

⌗ **To begin this exercise, undo the rotate copy process of the former exercise so that you are left with a single arc. Rotate it and move it to the left so that you have a bowl-shaped arc placed left of the center of your screen, as in *Figure 5.10.***

Except where noted, you should have snap and ortho on to do this exercise.

⌗ **Select the arc.**

⌗ **Type "Mi", select Mirror from the Draw menu, or the Mirror tool from the Modify toolbar, as shown in *Figure 5.11.***

Now AutoCAD will ask you for the first point of a mirror line.

First point of mirror line:

A mirror line is just what you would expect; the line serves as the "mirror" and all points on your original object will be reflected across the line at an equal distance and opposite orientation.

We will show a mirror line even with the top of the arc, so that the end points of the mirror images will be touching.

Figure 5.10.

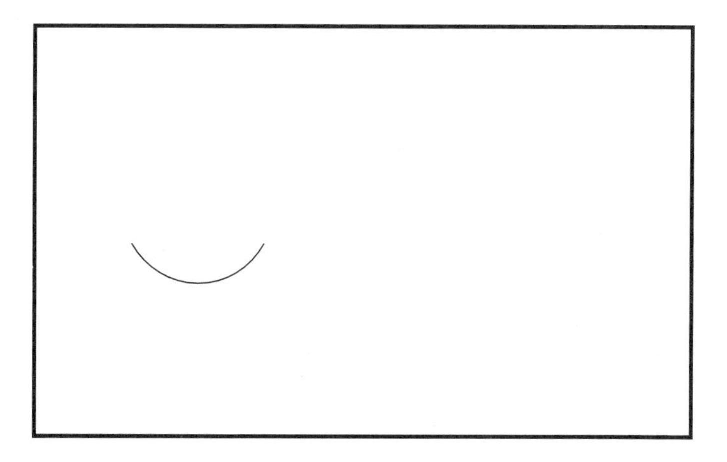

Figure 5.11.

⌖ **Select a point even with the left end point of the arc, as in *Figure 5.12.***

You are prompted to show the other end point of the mirror line:

Second point:

The length of the mirror line is not important. All that matters is its orientation. Move the cursor slowly in a circle, and you will see an inverted copy of the arc moving with you to show the different mirror images that are possible, given the first point you have specified. Turn ortho off to see the whole range of possibilities, then turn it on again to complete the exercise.

We will select a point at 0 degrees from the first point, so that the mirror image will be directly above the original arc, and touching at the end points as in *Figure 5.12.*

⌖ **Select a point directly to the right (0 degrees) of the first point.**

The dragged object will disappear until you answer the next prompt, which asks if you want to delete the original object or not.

Delete old objects? <N>:

This time around we will not delete the original.

⌖ **Press Enter to retain the old object.**

Your screen will look like *Figure 5.12,* without the mirror line in the middle.

Mirroring with Grips

The Mirror grip edit mode works exactly like the rotate mode, except that the rubber band will show you a mirror line instead of a rotation angle. The option to retain or delete the original is obtained through the copy option, just as in the Rotate mode. Try it.

⌖ **Select the two arcs on your screen by pointing or windowing.**

Figure 5.12.

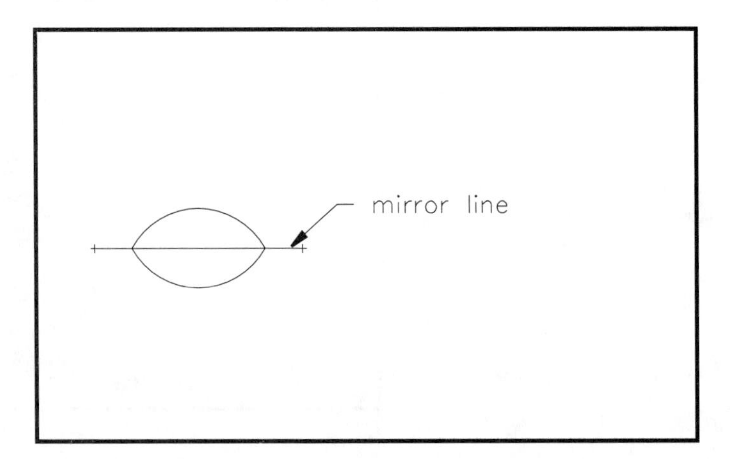

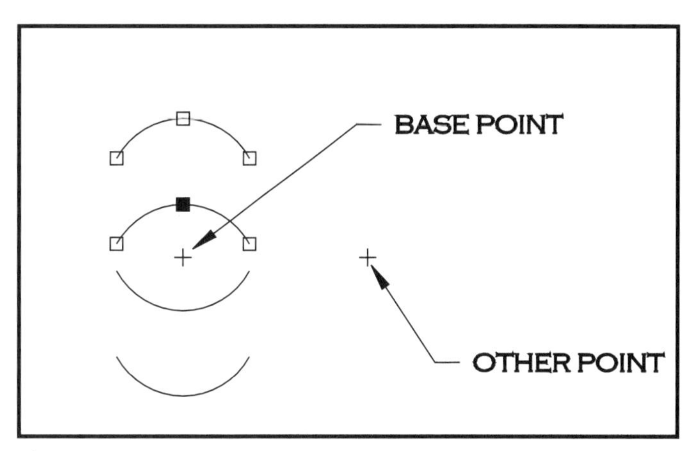

Figure 5.13.

The arcs will be highlighted and grips will be showing.

Pick any of the grips.

Press the enter button on your mouse and pick Mirror from the pop up menu.

Move the cursor and observe the dragged mirror images of the arcs. Notice that the rubber band operates as a mirror line, just as in the Mirror command.

Type "b" or press the enter button again and select "Base point".

This frees you from the selected grip and allows you to create a mirror line from any point on the screen. Notice the "Base point:" prompt in the command area.

Type "c" or open the cursor menu and select "Copy".

As in the Rotate mode, this is how you retain the original in a grip edit mirroring sequence.

Pick a base point slightly below the arcs.

Pick a second point to the right of the first.

Your screen should resemble *Figure 5.13.*

Press enter or the space bar to exit grip edit mode.

5.5 CHANGING PAPER SIZE, ORIENTATION, PLOT SCALE, ROTATION, AND ORIGIN

This discussion will complete our exploration of the basic features of the Plot Configuration dialog box. The parameters involved will dramatically impact the results of your hard copy output. In particular, putting all of these variables together with each other and previously discussed parameters takes a lot of skill and knowhow. You can gain this expertise through experience with actual plotting and plot previewing. As you become more comfortable with what is going on, it is especially valuable experience to plot using more than one device and different paper sizes.

Type Ctrl-P or select the Print tool from the standard tool bar to open the Plot Configuration dialog box.

Paper Size and Orientation

This box gives critical information and choices about the size and orientation of your drawing sheet. To begin with, you can choose to see information presented in inches or millimeters by selecting one of the two radio buttons. Inches is the default.

Click on "MM".

You will see that both the plot area and the scale values are changed to reflect metric units.

Click on "Inches".

On the right you will see a rectangle showing the orientation of the paper in the plotting device. It will be either landscape (horizontal) or portrait (vertical). This orientation is part of the device driver and cannot be changed directly. If you want to plot horizontally when your device is configured vertically, or vice-versa, you must rotate the plot using the Rotation and Origin dialog described in the next section.

The plot area is shown below the radio buttons and is also determined by the plotting device. The default size will be the maximum size available on your printer or plotter. If you are using a printer, for example, you may have the equivalent of an A-size sheet. In this case the plot area will be close to 8.00 × 11.00.

Click on "Size".

This will open the Paper Size dialog box shown in *Figure 5.14.* (If you are using a printer with only one paper size, this dialog box may be inaccessible.) The standard sizes listed in the box on the left are device specific and will depend on your plotter. The "USER" boxes on the right allow you to define plotting areas of your own. This means, for example, that you can plot in a 5.00 × 5.00 area on an 8.00 × 11.00 or larger sheet of paper.

Paper size can be changed by picking any of the sizes on the list at the left.

Change the paper size, if you wish, and then click on "OK" to exit the dialog.

If you have changed sizes, the new size will be named next to the Size box, and the new plot area will be listed below. The change also may be reflected in the scale box ("Plotted Inches = Drawing Units").

If you have changed paper sizes we suggest that you do a plot preview. Paper size is one of the important factors in determining effective plotting area.

Scale, Rotation, and Origin

Plots can be scaled to fit the available plot area, or given an explicit scale of paper units to drawing units. "Scaled to Fit" is the default, as shown by the check in the check box.

Figure 5.14.

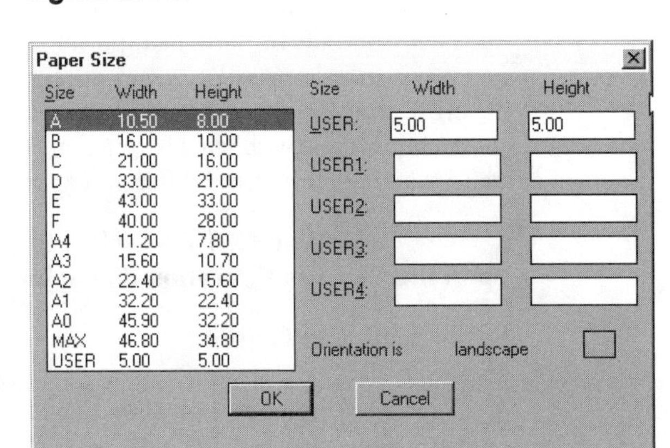

With this setting, the area chosen for plotting in the Additional Parameters box (Display, Extents, Limits, etc.) will be plotted as large as possible within the plot area specified in the Paper Size and Orientation box. Fitting the chosen area to the available plot area will dictate the scale shown in the edit boxes. Plotted inches or millimeters are shown in relation to drawing units. If either the area to be plotted or the paper size is changed, the change will be reflected in the scale boxes.

If "Scaled to Fit" is not checked, the scale will default to 1 = 1. In this case drawing units will be considered equivalent to paper size units. 1-to-1 scale is definitely preferred when plotting from Paper Space (Chapter 6). Other scales can be specified explicitly by typing in the edit boxes. When you change the drawing scale, the area to be plotted may be affected. If the area is too large for the paper size given the scale, then only a portion of the chosen area will be plotted. If the area becomes smaller than the available paper, some blank space will be left. Changing scales should be followed by a partial preview.

The area of the paper that is actually used for plotting may also be affected by the rotation and origin of the plot. If your device is configured with paper oriented vertically (portrait style), and you want to plot horizontally (landscape), then you will need to rotate 90 degrees. This is done with the radio buttons in the Plot Rotation and Origin dialog box.

⌗ **Click in the Rotation and Origin box.**

This will call the dialog box shown in *Figure 5.15*. You can rotate to 0, 90, 180, or 270 degrees as shown. Rotating a plot by 90 degrees will have a very significant impact on the relationship between paper size and effective drawing area.

This box will also allow you to change the origin of the plot. Plots usually originate in the lower left-hand corner of the page, but you can alter this. For example, let's say you wanted to plot a 5 × 5 area in the upper right of an 8 × 11 sheet, landscape orientation. You could do this by creating a 5 × 5 "User" paper size and then moving its origin from (0,0) to (6,3).

⌗ **Change the plot rotation if you wish, and then Click on OK to exit the dialog box.**

At this point, you have been through all the major parameters of plot configuration. It is important that you gain experience in using the configuration parameters available. When you have a drawing ready, open the plot configuration dialog box and make whatever adjustments you think are necessary. You should access at least one partial and one full preview along the way. When everything looks right, check your plotting device and paper and then click on "OK" to start it rolling.

Figure 5.15.

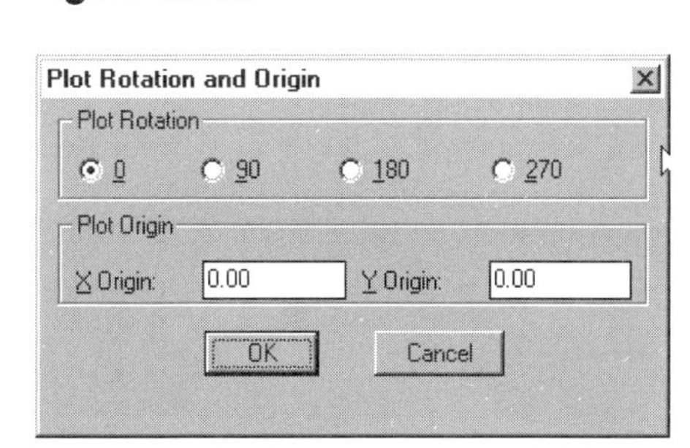

5.6 REVIEW MATERIAL

Questions:

1. What factors define a polar array? How many are needed to define an array?
2. What factors define an arc? How many are needed for any single method?
3. How would you use the reference option to rotate a line from 60 degrees to 90 degrees? How would you accomplish the same rotation without using reference?
4. What is the purpose of the base point option in the grip edit Rotate mode?
5. Why does Mirror require a mirror line where Rotate only requires a single point?
6. How do you create a landscape plot if your printer prints in portrait orientation?
7. How would you print within a 3 × 3 area in the lower left corner of an 8.5 × 11 peice of paper?

COMMANDS

Draw	Modify
ARC	ARRAY (polar)
	MIRROR
	ROTATE

Drawing Problems

1. Draw an arc starting at (10,6) and passing through (12,6.5) and (14,6).
2. Create a mirrored copy of the arc across the horizontal line passing through (10,6).
3. Rotate the pair of arcs from problem two 45 degrees around the point (9,6).
4. Create a mirrored copy of the pair of arcs mirrored across a vertical line passing through (9,6).
5. Create mirrored copies of both pairs of arcs mirrored across a horizontal line passing through (9,6).
6. Erase any three of the four pairs of arcs on your screen and recreate them using a polar array.

DRAWING 5—1

Dials

This is a relatively simple drawing that will give you some good practice with polar arrays and the ROTATE and COPY commands.

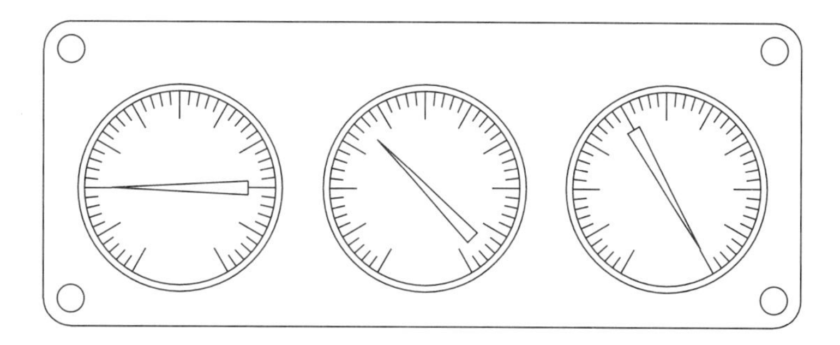

Notice that the needle drawn at the top of the next page is only for reference; the actual drawing includes only the plate and the three dials with their needles.

Drawing Suggestions

GRID = .25   LTSCALE = .50
SNAP = .125   LIMITS = (0,0) (12,9)

- After drawing the outer rectangle and screw holes, draw the leftmost dial, including the needle. Draw a .50 vertical line at the top and array it to the left (counterclockwise—a positive angle) and to the right (negative) to create the 11 larger lines on the dial. Use the same operation to create the 40 small (.25) markings.
- Complete the first dial and then use a multiple copy to produce two more dials at the center and right of your screen. Be sure to use a window to select the entire dial.
- Finally, use the ROTATE command to rotate the needles as indicated on the new dials. Use a window to select the needle, and rotate it around the center of the dial.

Detail of pointer

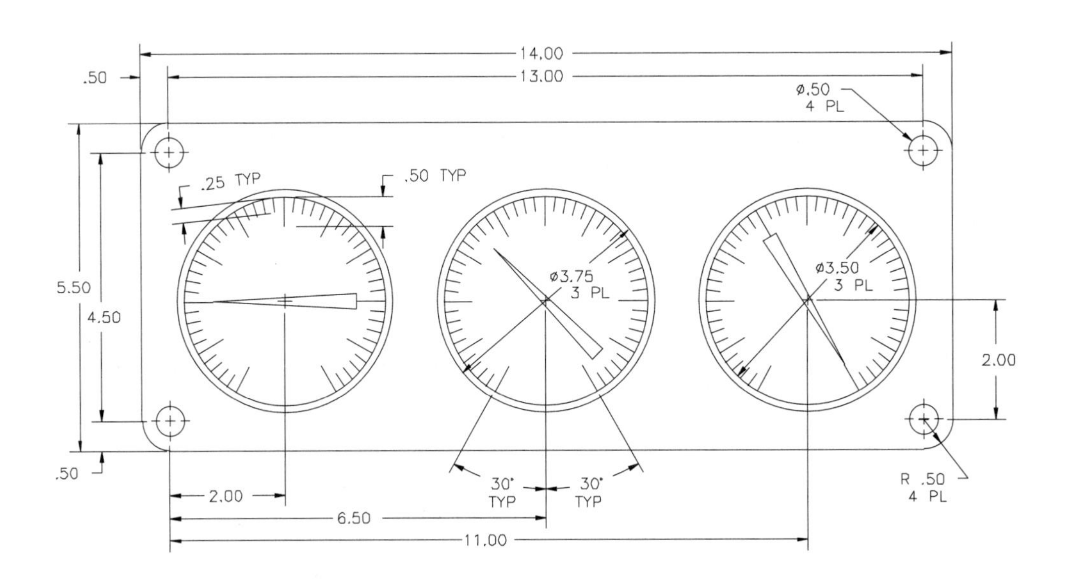

DIALS

Drawing 5–1

DRAWING 5—2

Alignment Wheel

This drawing shows a typical use of the MIRROR command. Carefully mirroring sides of the symmetrical front view will save you from duplicating some of your drawing efforts. Notice that you will need a small snap setting to draw the vertical lines at the chamfer.

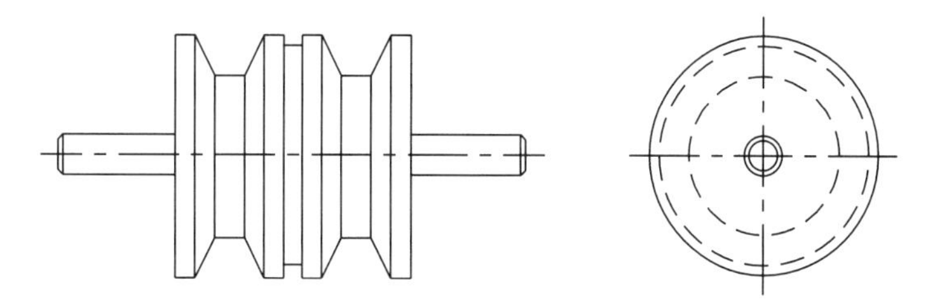

Drawing Suggestions

GRID = .25 LTSCALE = .50
SNAP = .125 LIMITS = (0,0) (12,9)

- There are numerous ways to use Mirror in drawing the front view. As the reference shows, there is top-bottom symmetry as well as left-right symmetry. The exercise for you is to choose an efficient mirroring sequence.
- Whatever sequence you use, consider the importance of creating the chamfer and the vertical line at the chamfer before this part of the object is mirrored.
- Once the front view is drawn, the right side view will be easy. Remember to change layers for center and hidden lines and to line up the small inner circle with the chamfer.

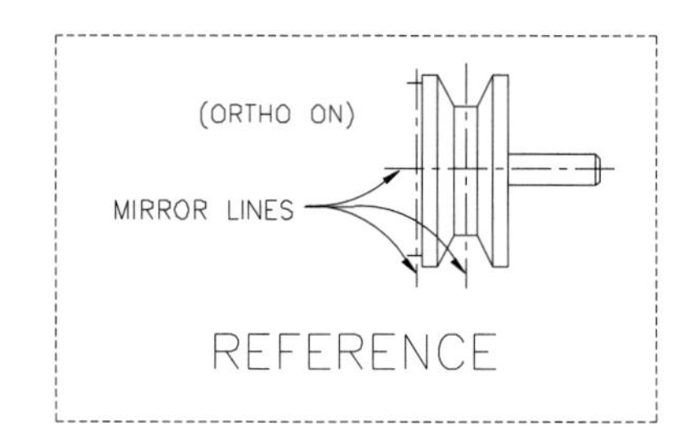

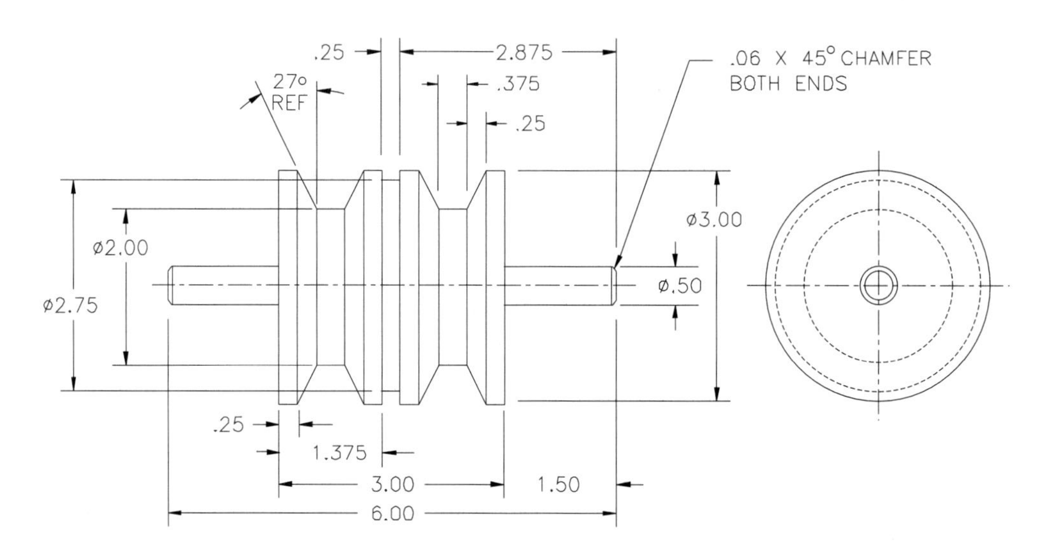

ALIGNMENT WHEEL
Drawing 5–2

DRAWING 5—3

Hearth

Once you have completed this architectural drawing as it is shown, you might want to experiment with filling in a pattern of firebrick in the center of the hearth. The drawing itself is not complicated, but little errors will become very noticeable when you try to make the row of 4 × 8 bricks across the bottom fit with the arc of bricks across the top, so work carefully.

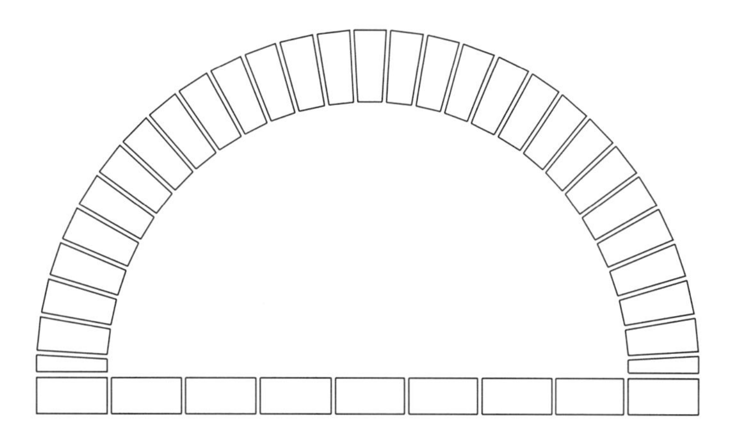

Drawing Suggestions

UNITS = Architectural smallest fraction = 8 (1/8")
LIMITS = (0,0) (12',9')
GRID = 1'
SNAP = 1/8"

- Zoom in to draw the wedge-shaped brick indicated by the arrow on the right of the dimensioned drawing. Draw half of the brick only and mirror it across the center line as shown. (Notice that the center line is for reference only.) It is very important that you use Mirror so that you can erase half of the brick later.
- Array the brick in a 29 item, 180 degree polar array.
- Erase the bottom halves of the end bricks at each end.
- Draw a new horizontal bottom line on each of the two end bricks.
- Draw a 4 × 8 brick directly below the half brick at the left end.
- Array the 4 × 8 brick in a 1 row, 9 column array, with 8.5" between columns.

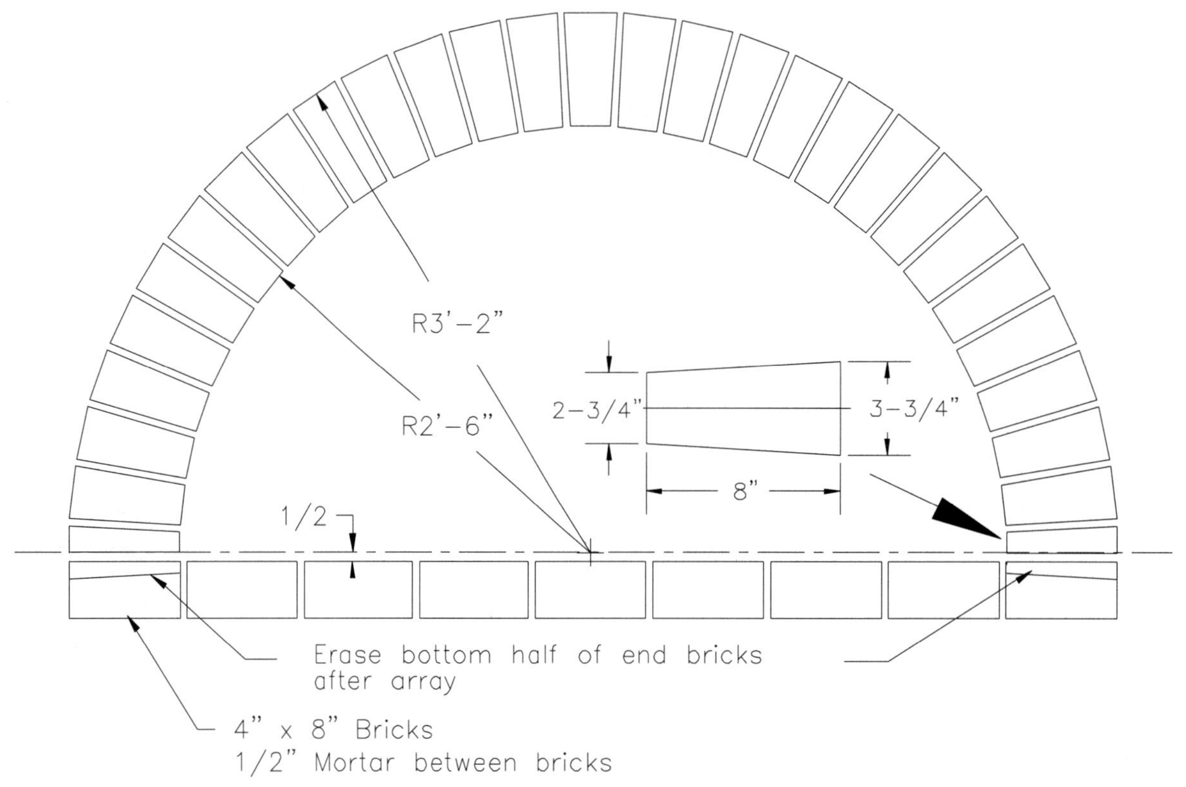

HEARTH

Drawing 5−3

6

Object Snap

OVERVIEW

This chapter will continue to expand your repertoire of editing commands. You will learn to BREAK entities on the screen into pieces so that they may be manipulated or erased separately. You will also learn to shorten entities using the TRIM command, or to lengthen them with the EXTEND command.

But most important, you will begin to use a very powerful tool called Object Snap that will take you to a new level of accuracy and efficiency as a CAD operator.

SECTIONS

OBJECTIVES

After reading this chapter, you should be able to:

- Select Points with Object Snap (Single-point Overrides).
- Select Points with OSNAP (Running Modes).
- BREAK Previously Drawn Objects.
- Shorten Objects with the TRIM Command.
- Extend Objects with the EXTEND Command.
- Draw an Archimedes spiral.
- Draw spiral designs.
- Draw a grooved hub.

6.1 SELECTING POINTS WITH OBJECT SNAP (SINGLE-POINT OVERRIDE)

Some of the drawings in the last two chapters have pushed the limits of what you can accomplish accurately on a CAD system with incremental snap alone. Object snap is a related tool that works in a very different manner. Instead of snapping to points defined by the coordinate system, it snaps to geometrically specifiable points on objects that you already have drawn.

Let's say you want to begin a new line at the end point of one that is already on the screen. If you are lucky, it may be on a snap point, but it is just as likely not to be. Turning snap off and moving the cursor to the apparent end point may appear to work, but when you zoom in you probably will find that you have missed the point. Using object snap is the only precise way, and it is as precise as you could want. Let's try it.

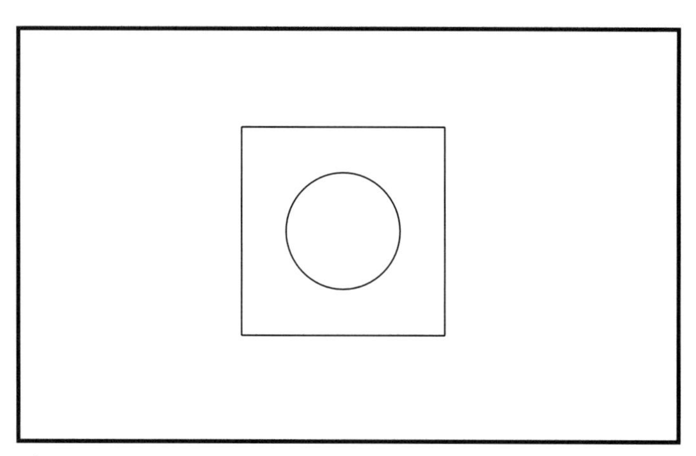

Figure 6.1.

⌗ To prepare for this exercise, draw a 6 x 6 box with a circle inside, as in *Figure 6.1.*

Exact sizes and locations are not important; however, the circle should be centered within the square.

⌗ Now enter the LINE command (type "L" or select the Line tool).

We are going to draw a line from the lower left corner of the square to a point on a line tangent to the circle, as shown in *Figure 6.2.* This task would be extremely difficult without object snap. The corner is easy to locate, since you probably have drawn it on snap, but the tangent may not be.

We will use an "endpoint" object snap to locate the corner and a "tangent" object snap to locate the tangent point. When AutoCAD asks for a point, you can type the abreviated name of an object snap mode, or select an object snap mode from the handy cursor menu.

⌗ At the "From point:" prompt, instead of specifying a point, type "end" or hold down the CTRL key and press the enter button on your cursor.

This will open the pop up cursor menu illustrated in *Figure 6.3.*

⌗ If you are using the cursor menu, select "Endpoint".

Figure 6.2.

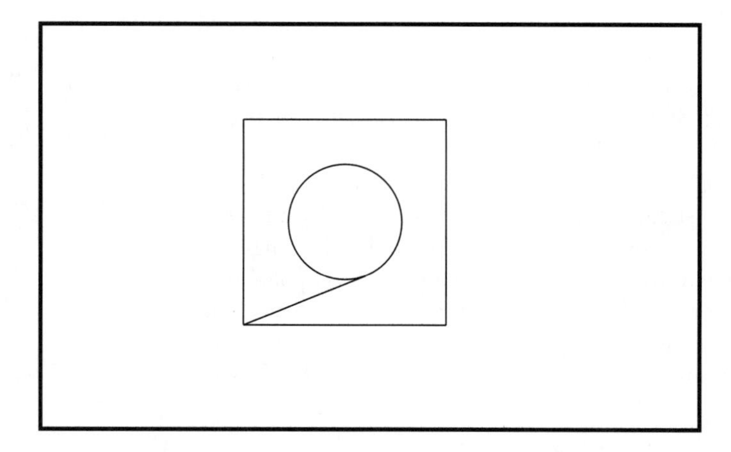

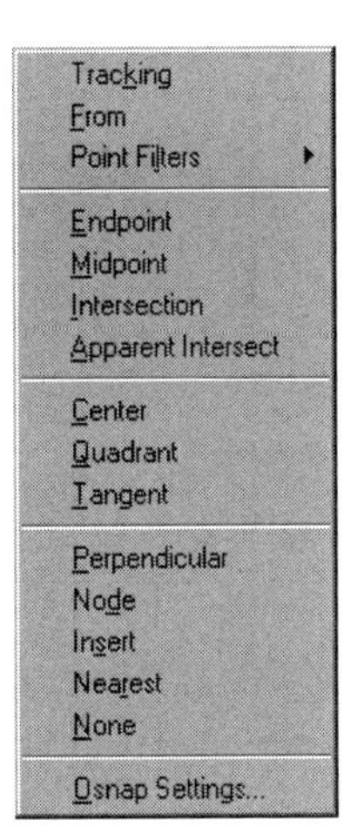

Figure 6.3.

Entering "Endpoint" by either of these methods tells AutoCAD that you are going to select the start point of the line by using an end point object snap rather than by direct pointing or entering coordinates.

Now that AutoCAD knows that we want to begin at the end point of a previously drawn entity, it needs to know which one.

⌖ **Move the cursor near the lower left corner of the square.**

When you are close to the corner, AutoCAD will recognize the end point of the lines there and indicate this with a yellow box surrounding the endpoint. This object snap symbol is called a marker. There are different shaped markers for each type of object snap. If you let the cursor rest here for a moment, a tooltip-like label will appear, naming the type of object that has been recognized, as shown in *Figure 6.4.* This label is called a snaptip. Also notice that if the cross hairs are inside the marker, they will lock onto the end point. this action can be turned on or off and is called the magnet setting. You do not have to be locked on to the end point, however. As long as the end point marker is showing the end point will be selected.

⌖ **With the end point object snap marker showing, press the pick button.**

The yellow end point box and the snaptip will disappear, and there will be a rubber band stretching from the lower left corner of the square to the cursor position. In the command area you will see the To point prompt.

We will use a tangent object snap to select the second point.

⌖ **At the "To point:" prompt, type "tan" or select "Tangent" from the cursor menu (shift + enter button to open the cursor menu).**

⌖ **Move the cursor to the right and position the cross hairs so that they are near the lower right side of the circle.**

When you approach the tangent area, you will see the yellow tangent marker, as shown in *Figure 6.5.* Here again, if you let the cursor rest you will see a snaptip.

⌖ **With the tangent marker showing, press the pick button.**

AutoCAD will locate the tangent point and draw the line.

⌖ **Press Enter to exit the LINE command.**

Your screen should now resemble *Figure 6.2.*

We will repeat the process now, but start from the midpoint of the bottom side of the square instead of its end point.

Figure 6.4.

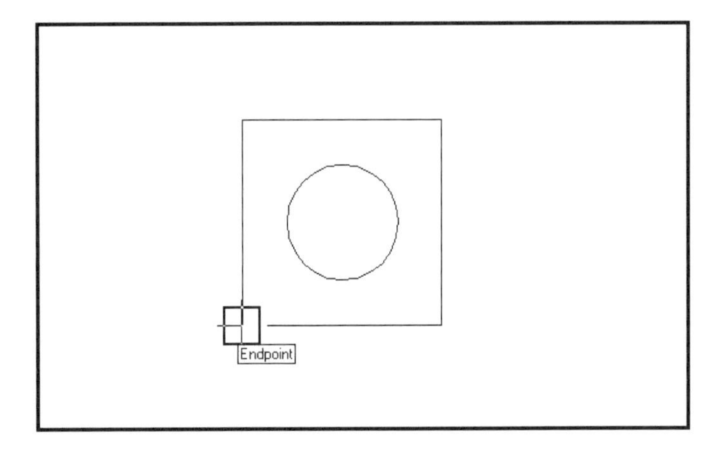

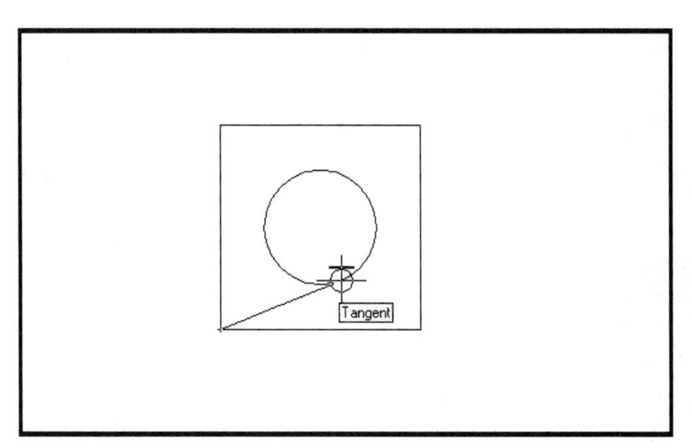

Figure 6.5.

⌖ **Repeat the LINE command.**

⌖ **At the prompt for a point, select "Midpoint" from the cursor menu.**

⌖ **Position the aperture anywhere along the bottom side of the square so that a yellow triangle, the midpoint marker, appears at the midpoint of the line.**

⌖ **Press the pick button.**

⌖ **At the prompt for a second point, select "Tangent" from the cursor menu.**

⌖ **Position the aperture along the right side of the circle so that the tangent symbol appears and press the pick button.**

⌖ **Press Enter or the space bar to exit the LINE command.**

At this point your screen should resemble *Figure 6.6.*

That's all there is to it. Remember the steps: 1) enter a command; 2) when AutoCAD asks for a point, type or select an object snap mode; 3) position the cross hairs near an object to which the mode can be applied and let AutoCAD find the point.

Figure 6.6.

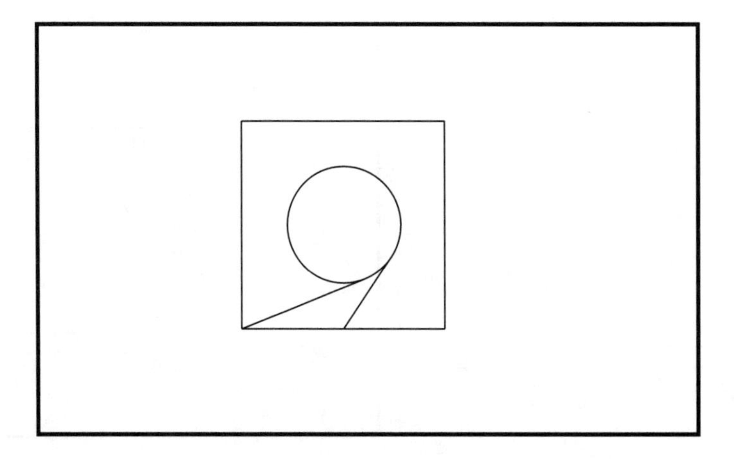

6.2 SELECTING POINTS WITH OSNAP (RUNNING MODE)

So far we have been using object snap one point at a time. Since object snap is not constantly in use for most applications, this single-point method is probably most common. But if you find that you are going to be using one or a number of object snap types repeatedly and will not need to select many points without them, there is a way to keep object snap modes on so that they affect all point selection. These are called "running object snap modes". We will use this method to complete the drawing shown in *Figure 6.7.* Notice how the lines are drawn from midpoints and corners to tangents to the circle. This is easily done with object snap.

⌖ **Type "os" or select "Object Snap Settings . . ." from the Tools menu.**

You will see the dialog box illustrated in *Figure 6.8.* At the top is a list of object snap modes with check boxes. At the bottom is a box with a scroll bar where you can change the object snap aperture size.

You will find a description of all of the object snap modes on the chart (Figure 6.9) at the end of this task, but for now we will be using three: midpoint, tangent, and intersect. Midpoint and tangent you already know. Intersect snaps to the point where two entities meet or cross. We will use intersect instead of end point to select the remaining three corners of the square.

⌖ **Click on the check boxes next to "Midpoint", "Tangent", and "Intersection".**

⌖ **Click OK.**

⌖ **Enter the LINE command.**

⌖ **Position the aperture so that the lower right corner is within the box.**

A yellow X, the intersection marker, will appear.

⌖ **With the intersection marker showing, press the pick button.**

AutoCAD will select the intersection of the bottom and the right sides and give you the rubber band and the prompt for a second point.

⌖ **Move the cross hairs up and along the right side of the circle until the tangent marker appears.**

⌖ **With the tangent marker showing, press the pick button.**

Figure 6.7.

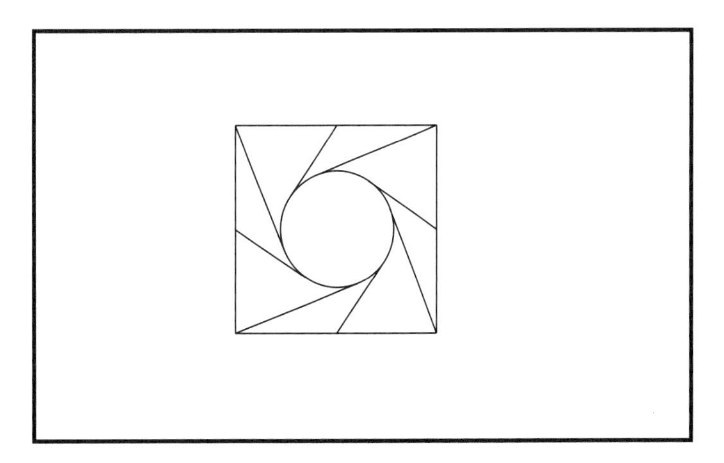

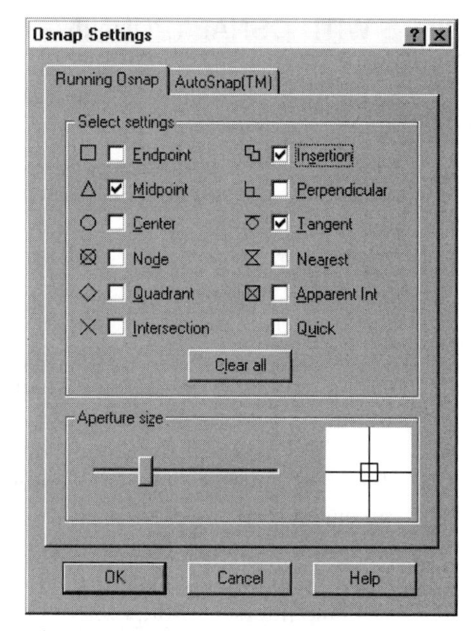

Figure 6.8.

AutoCAD will construct a new tangent from the lower right corner to the circle.

Press Enter to complete the command sequence.

Press Enter again to repeat LINE so you can begin with a new start point.

We will continue to move counterclockwise around the circle. This should begin to be easy now.

Position the aperture along the right side of the square so that the midpoint triangle appears.

With the midpoint marker showing, press the pick button.

AutoCAD snaps to the midpoint of the side.

Move up along the upper right side of the circle so that the tangent marker appears.

With the tangent marker showing, press the pick button.

Press Enter to exit LINE.

Press Enter again to repeat LINE and continue around the circle drawing tangents like this: upper right corner to top of circle, top side midpoint to top left of circle, upper left corner to left side, left side midpoint to lower left side.

Remember that running osnap modes should give you both speed and accuracy, so push yourself a little to see how quickly you can complete the figure.

Your screen should now resemble *Figure 6.7.*

Before going on we need to turn off the running osnap modes.

Type "os" or select "Object Snap Settings..." from the Tools menu.

Click the Clear All box.

Click on OK to exit the dialog box.

TIP: One of the object snap modes is called None, meaning no object snap. F3 is the function key equivalent of None. This can be used while running object snaps are in

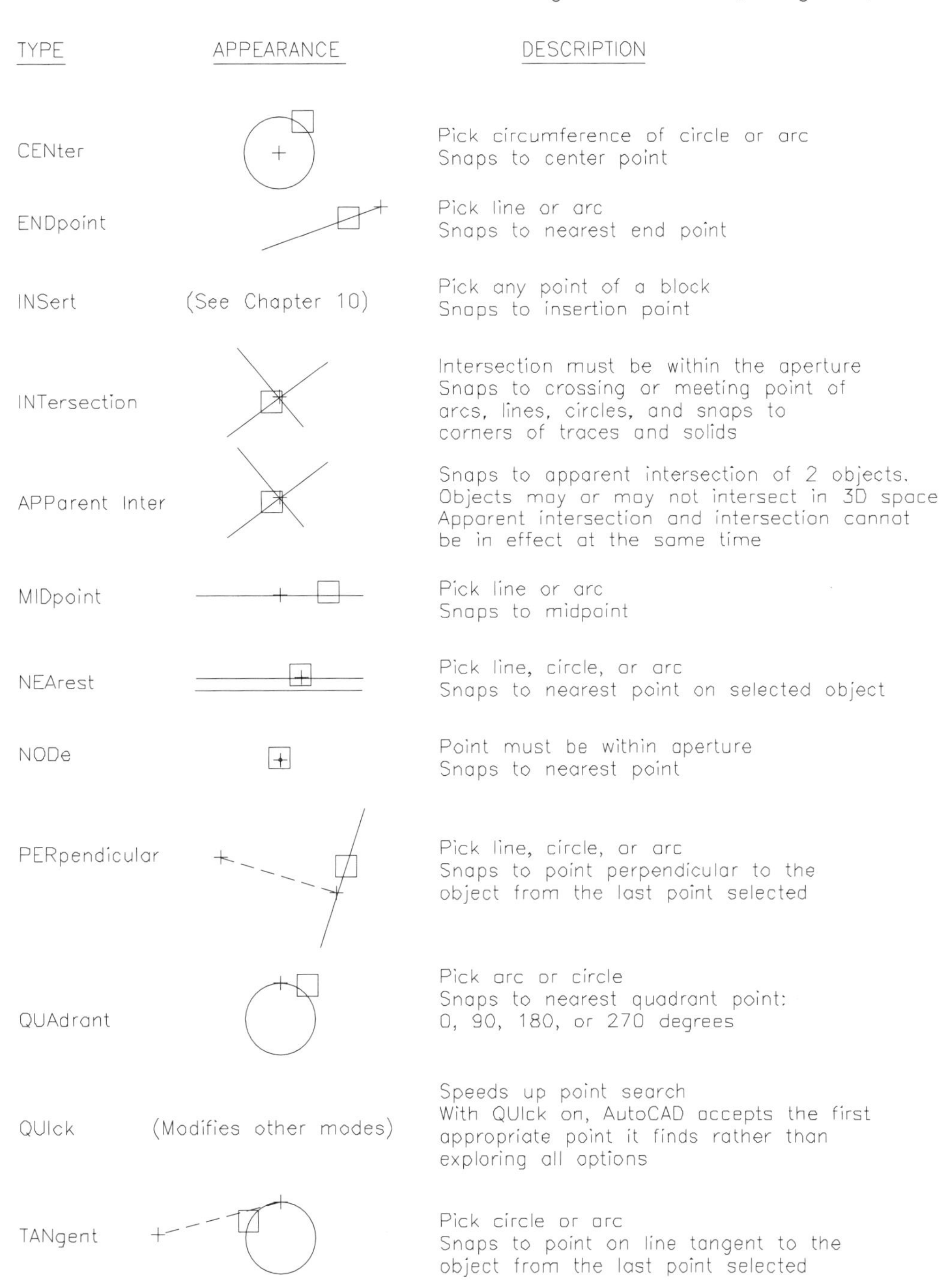
TYPE
APPEARANCE
DESCRIPTION
CENter
Pick circumference of circle or arc
Snaps to center point
ENDpoint
Pick line or arc
Snaps to nearest end point
INSert
(See Chapter 10)
Pick any point of a block
Snaps to insertion point
INTersection
Intersection must be within the aperture
Snaps to crossing or meeting point of
arcs, lines, circles, and snaps to
corners of traces and solids
APParent Inter
Snaps to apparent intersection of 2 objects.
Objects may or may not intersect in 3D space
Apparent intersection and intersection cannot
be in effect at the same time
MIDpoint
Pick line or arc
Snaps to midpoint
NEArest
Pick line, circle, or arc
Snaps to nearest point on selected object
NODe
Point must be within aperture
Snaps to nearest point
PERpendicular
Pick line, circle, or arc
Snaps to point perpendicular to the
object from the last point selected
QUAdrant
Pick arc or circle
Snaps to nearest quadrant point:
0, 90, 180, or 270 degrees
QUIck
(Modifies other modes)
Speeds up point search
With QUIck on, AutoCAD accepts the first
appropriate point it finds rather than
exploring all options
TANgent
Pick circle or arc
Snaps to point on line tangent to the
object from the last point selected

Figure 6.9.

effect and you wish to select a point that the object snap modes would interfere with. None or F3 work like a single point override in reverse. They override the running object snaps for a single point or selection.

Now we will move on to five very useful and important new editing commands: BREAK, TRIM, EXTEND, STRETCH, and LENGTHEN. Before leaving Object Snap, be sure to study the chart, *Figure 6.9*.

6.3 BREAKING PREVIOUSLY DRAWN OBJECTS

The BREAK command allows you to break an object on the screen into two entities, or to cut a segment out of the middle or off the end. The command sequence is similar for all options. The action taken will depend on the points you select for breaking.

⌖ In preparation for this section, clear your screen of any objects left over from Task 2 and draw a 5.0 horizontal line across the middle of your screen, as in *Figure 6.10*.

Exact lengths and coordinates are not important. Also, be sure to turn off any running object snap modes that may be on from the last exercise.

AutoCAD allows for four different ways to break an object, depending on whether the point you use to select the object is also to be considered a break point. You can break an object at one point or at two points, and you have the choice of using your object selection point as a break point or not.

We begin by breaking the line you have just drawn into two independent lines using a single break point, which is also the point used to select the line.

⌖ Type "Break" or select the Break tool from the Modify menu.

There is also a Break tool on the Modify toolbar, as shown in *Figure 6.11*, but it is at the bottom of the toolbar in the docked position and may be hidden behind the command prompt. If this is the case you can access it by moving the toolbar to a floating position.

Be aware that the noun/verb or pick first sequence does not work with BREAK.

Figure 6.10.

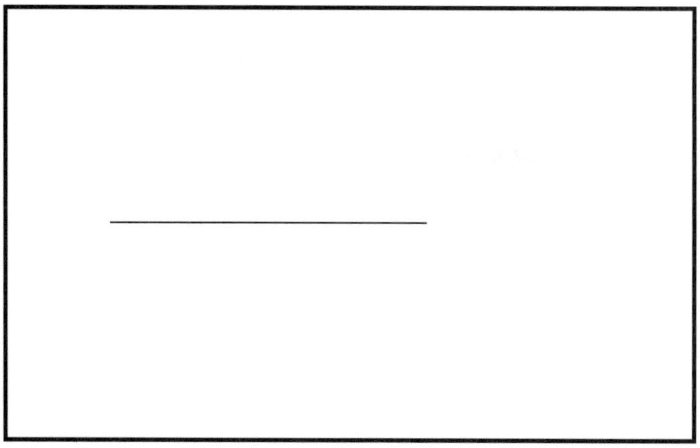

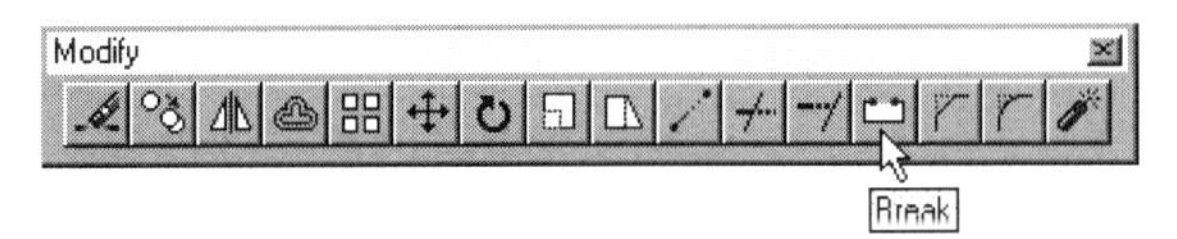

Figure 6.11.

AutoCAD will prompt you to select an object to break:

Select object:

You may select an object in any of the usual ways, but notice that you can only break one object at a time. If you try to select more—with a window, for example—AutoCAD will give you only one. Because of this you will best indicate the object you want to break by pointing to it.

TIP: Object snap modes work well in edit commands such as BREAK. If you wish to break a line at its midpoint, for example, you can use the midpoint object snap mode to select the line and the break point.

⌗ Select the line by picking any point near its middle. (The exact point is not critical; if it were, we could use a midpoint object snap.)

The line has now been selected for breaking, and since there can be only one object, you do not have to press Enter to end the selection process as you often do in other editing commands.

The break is complete. In order to demonstrate that the line is really two lines now, we will select the right half of it for our next break.

⌗ Press Enter or the space bar to repeat the BREAK command.

⌗ Point to the line on the right side of the last break.

The right side of the line should become dotted, as in *Figure 6.12.* Clearly the original line is now being treated as two separate entities.

We will shorten the end of this dotted section of the line. Assume that the point you just used to select the object is the point where you want it to end; now all you need to do is to select a second point anywhere beyond the right end of the line.

⌗ Select a second point beyond the right end of the line.

Your line should now be shortened, as in *Figure 6.13.*

Figure 6.12.

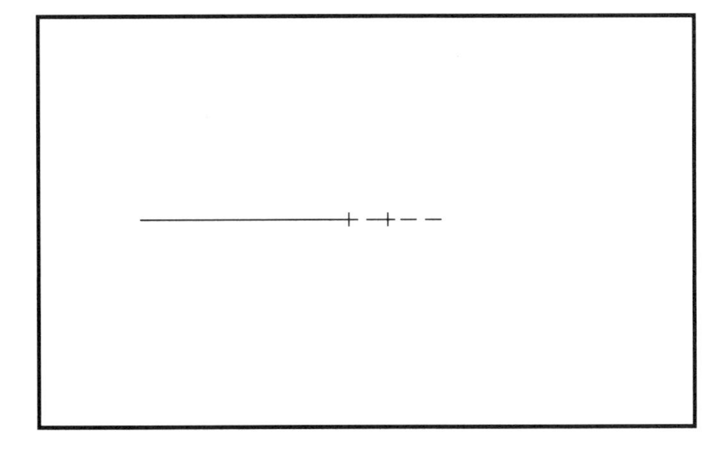

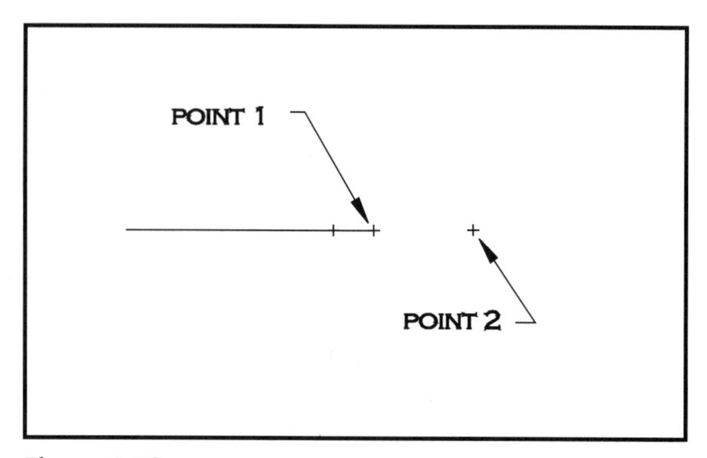

Figure 6.13.

BREAK is a very useful command, but there are times when it is cumbersome to shorten objects one at a time. The TRIM command has some limitations that BREAK does not have, but it is much more efficient in situations where you want to shorten objects at intersections.

6.4 SHORTENING OBJECTS WITH THE TRIM COMMAND

The TRIM command works wonders in many situations where you want to shorten objects at their intersections with other objects. It will work with lines, circles, arcs, and polylines (see Chapter 9). The only limitation is that you must have at least two objects and they must cross or meet. If you are not trimming to an intersection, use BREAK.

In preparation for exploring TRIM, clear your screen and then draw two horizontal lines crossing a circle, as in *Figure 6.14.* Exact locations and sizes are not important.

First we will use the TRIM command to go from *Figure 6.14* to *Figure 6.15.*

Figure 6.14.

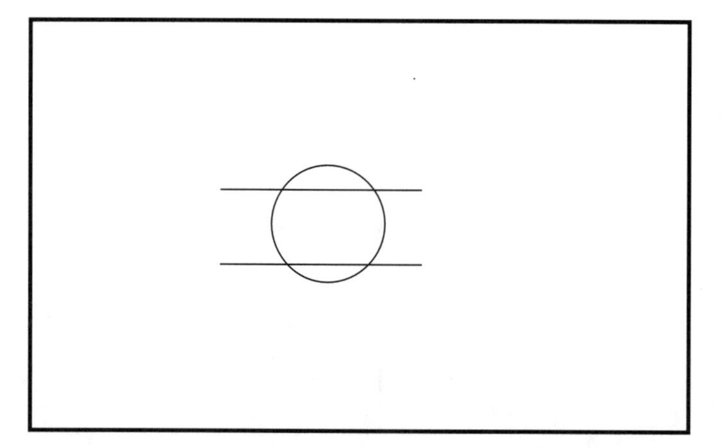

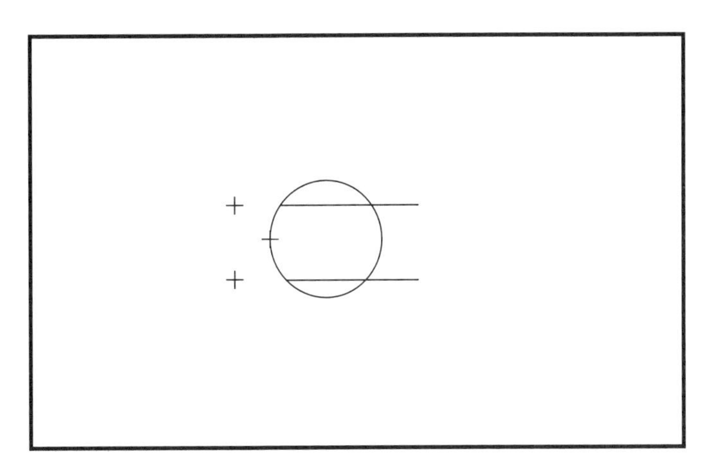

Figure 6.15.

⌗ **Type “Tr” or select Trim from the Modify menu, or the Trim tool from the Modify toolbar, as shown in Figure 6.16.**

The first thing AutoCAD will want you to specify is at least one cutting edge. A cutting edge is an entity you want to use to trim another entity. That is, you want the trimmed entity to end at its intersection with the cutting edge.

Select cutting edge(s): (Projmode = UCS, Edgemode = No extend)
Select objects:

The first line reminds you that you are selecting edges first—the objects you want to trim will be selected later. The option of selecting more than one edge is a useful one, which we will get to shortly. The notations in parentheses are relevant to 3D drawing and need not concern you at this point. For your information, Projmode and Edgemode are system variables that determine the way AutoCAD will interpret boundaries and intersections in 3D space.

For now we will select the circle as an edge and use it to trim the upper line.

⌗ **Point to the circle.**

The circle becomes dotted and will remain so until you leave the TRIM command. AutoCAD will prompt for more objects until you indicate that you are through selecting edges.

⌗ **Press Enter or the space bar to end the selection of cutting edges.**

You will be prompted for an object to trim:

<Select object to trim>/Project/Edge/Undo:

We will trim off the segment of the upper line that lies outside the circle on the left. The important thing is to point to the part of the object you want to remove, as shown in *Figure 6.15*.

⌗ **Point to the upper line to the left of where it crosses the circle.**

Figure 6.16.

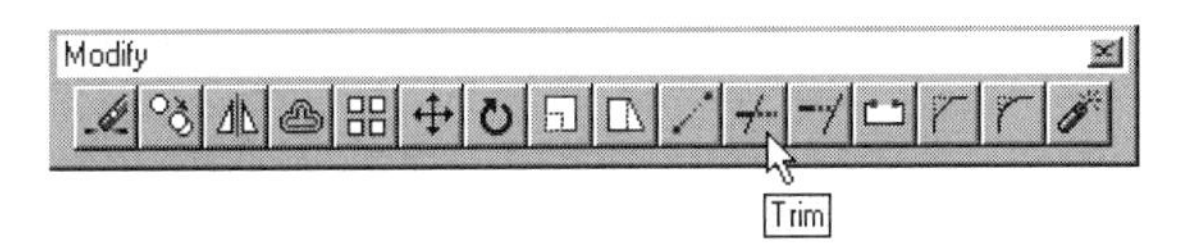

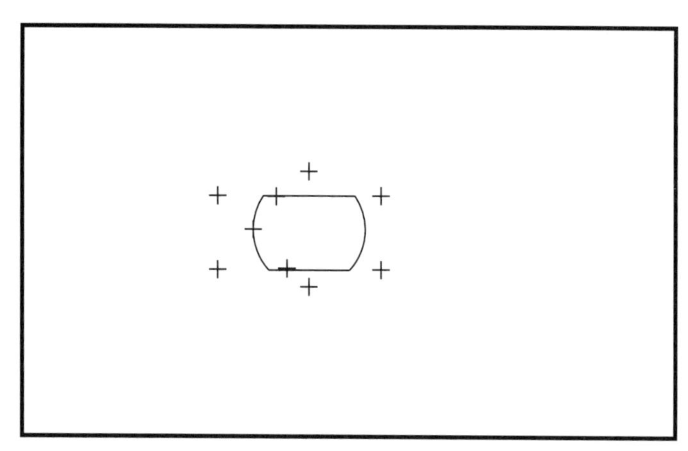

Figure 6.17.

The line is trimmed immediately, but the circle is still dotted, and AutoCAD continues to prompt for more objects to trim. Note how this differs from the BREAK command, in which you could only break one object at a time.

Also notice that you have an undo option, so that if the trim does not turn out the way you wanted, you can back up without having to leave the command and start over.

⌗ **Point to the lower line to the left of where it crosses the circle.**

Now you have trimmed both lines.

⌗ **Press Enter or the space bar to end the TRIM operation.**

Your screen should resemble *Figure 6.15.*

This has been a very simple trimming process, but more complex trimming is just as easy. The key is that you can select as many edges as you like and that an entity may be selected as both an edge and an object to trim, as we will demonstrate.

⌗ **Repeat the TRIM command.**

⌗ **Select both lines and the circle as cutting edges.**

This can be done with a window or with a crossing box.

⌗ **Press Enter to end the selection of edges.**

⌗ **Point to each of the remaining two line segments that lie outside the circle on the right, and to the top and bottom arcs of the circle to produce the band-aid shaped object in *Figure 6.17.***

⌗ **Press Enter to exit the TRIM command.**

6.5 EXTENDING OBJECTS WITH THE EXTEND COMMAND

If you compare the procedures of the EXTEND command and the TRIM command you will notice a remarkable similarity. Just substitute the word "boundary" for "cutting edge" and the word "extend" for "trim" and you've got it. These two commands are so quick to use that it is sometimes efficient to draw a temporary cutting edge or boundary on your screen and erase it after trimming or extending.

⌗ **Leave *Figure 6.17*, the band-aid, on your screen and draw a vertical line to the right of it, as in *Figure 6.18*.**

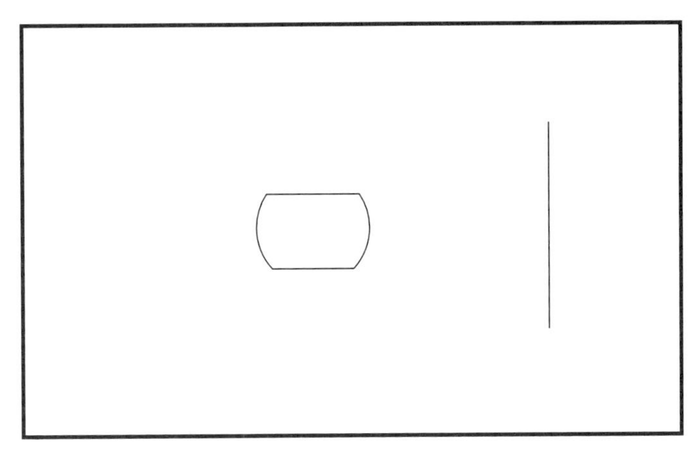

Figure 6.18.

We will use this line as a boundary to which to extend the two horizontal lines, as in *Figure 6.19.*

⊕ **Type "Ex" or select Extend from the Modify menu or the Extend tool from the Modify toolbar, as shown in *Figure 6.20.***

You will be prompted for objects to serve as boundaries:

Select boundary edge(s): (Projmode = UCS, Edgemode = No extend)
Select objects:

Look familiar? As with the TRIM command, any of the usual selection methods will work. For our purposes, simply point to the vertical line.

⊕ **Point to the vertical line on the right.**

You will be prompted for more boundary objects until you press Enter.

⊕ **Press Enter, the enter button, or the space bar to end the selection of boundaries.**

AutoCAD now asks for objects to extend:

<Select objects to extend>/Project/Edge/Undo:

⊕ **Point to the right half of one of the two horizontal lines.**

Figure 6.19.

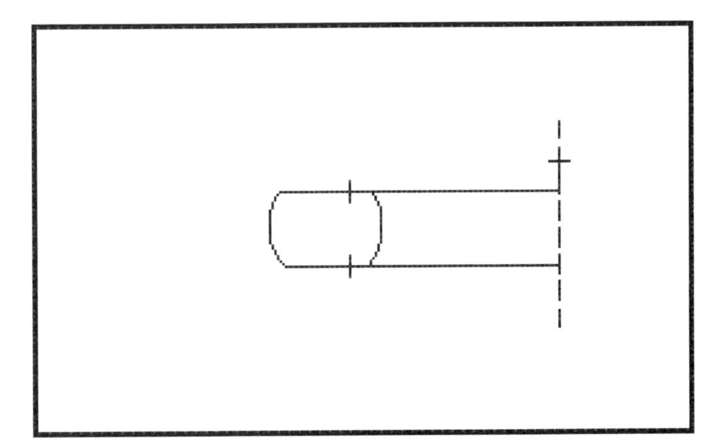

Figure 6.20.

Notice that you have to point to the line on the side closer to the selected boundary. Otherwise AutoCAD will look to the left instead of the right and give you the following message:

Object does not intersect an Edge

Note also that you only can select objects to extend by pointing. Windowing, crossing, or last selections will not work. Also be aware that arcs and polylines can be extended in the same manner as lines.

⌗ Point to the right half of the other horizontal line. Both lines should be extended to the vertical line.

Your screen should resemble *Figure 6.19.*

⌗ Press Enter to exit the EXTEND command.

6.6 REVIEW MATERIAL

Questions

1. What is an object snap marker? Give three examples.
2. At what point in a command procedure would you use an object snap single point override? How would you signal the AutoCAD program that you want to use an object snap?
3. How do you access the Object Snap cursor menu?
4. Why is there an object snap mode called "None "? What is its function key equivalent?
5. How do you use BREAK to shorten a line at one end? When would you use this procedure instead of the TRIM command?

COMMANDS

Modify	Tools
BREAK	OSNAP
TRIM	
EXTEND	

Drawing Problems

1. Draw a line from (6,2) to (11,6). Draw a second line perpendicular to the first starting at (6,6).
2. Break the first line at its intersection with the second.
3. There are now three lines on the screen. Draw a circle centered at their intersection and passing through the midpoint of the line going up and to the right of the intersection.
4. Trim all the lines to the circumference of the circle.
5. Erase what is left of the line to the right of the intersction and trim the portion of the circle to the left, between the two remaining lines.

DRAWING 6—1

Archimedes Spiral

This drawing and the next go together as an exercise you should find interesting and enjoyable. These are not technical drawings, but they will give you valuable experience with important CAD commands. You will be creating a spiral using a radial grid of circles and lines as a guide. Once the spiral is done, you will use it to create the designs in the next drawing, 6—3, "Spiral Designs."

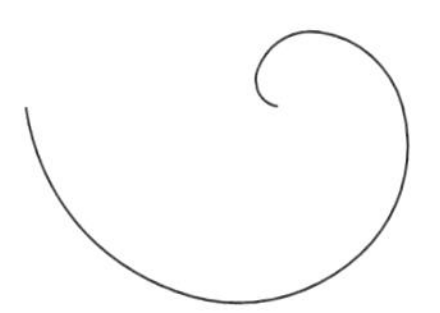

Drawing Suggestions

GRID = .5 LIMITS = (0,0) (18,12)

SNAP = .25 LTSCALE = .5

- The alternating continuous and hidden lines work as a drawing aid. If you use different colors and layers they will be even more helpful.
- Begin by drawing all the continuous circles on layer 0, centered near the middle of your display. Use the continuous circle radii as listed.
- Draw the continuous horizontal line across the middle of your six circles and then array it in a three-item polar array.
- Set to layer 2 for the hidden lines. The procedure for the hidden lines and circles will be the same as for the continuous lines, except the radii are different and you will array a vertical line instead of a horizontal one.
- Set to layer 1 for the spiral itself.
- Turn on a running object snap to intersection mode and construct a series of three-point arcs. Start points and end points will be on continuous line intersections; second points always will fall on hidden line intersections.
- When the spiral is complete, turn off layers 0 and 2. There should be nothing left on your screen but the spiral itself. Save it or go on to Drawing 6—2.

GROUPing objects

Here is a good opportunity to use Release 14's GROUP command. GROUP is discussed more fully in Chapter 10, but you will find it useful here. GROUP defines a collection of objects as a single entity so that they may be selected and modified as a unit. The spiral you have just drawn will be used in the next drawing and it will be easier to manipulate if you GROUP it. For more information, see Chapter 10.

1. Type "Group" or select the Object Group tool from the Standard toolbar.
2. Type "Spiral" for a group name.
3. Click on "New <".
4. Select the six arcs with a window.
5. Press Enter to end selection.
6. Click on OK.

The spiral can now be selected, moved, rotated, and copied as a single entity.

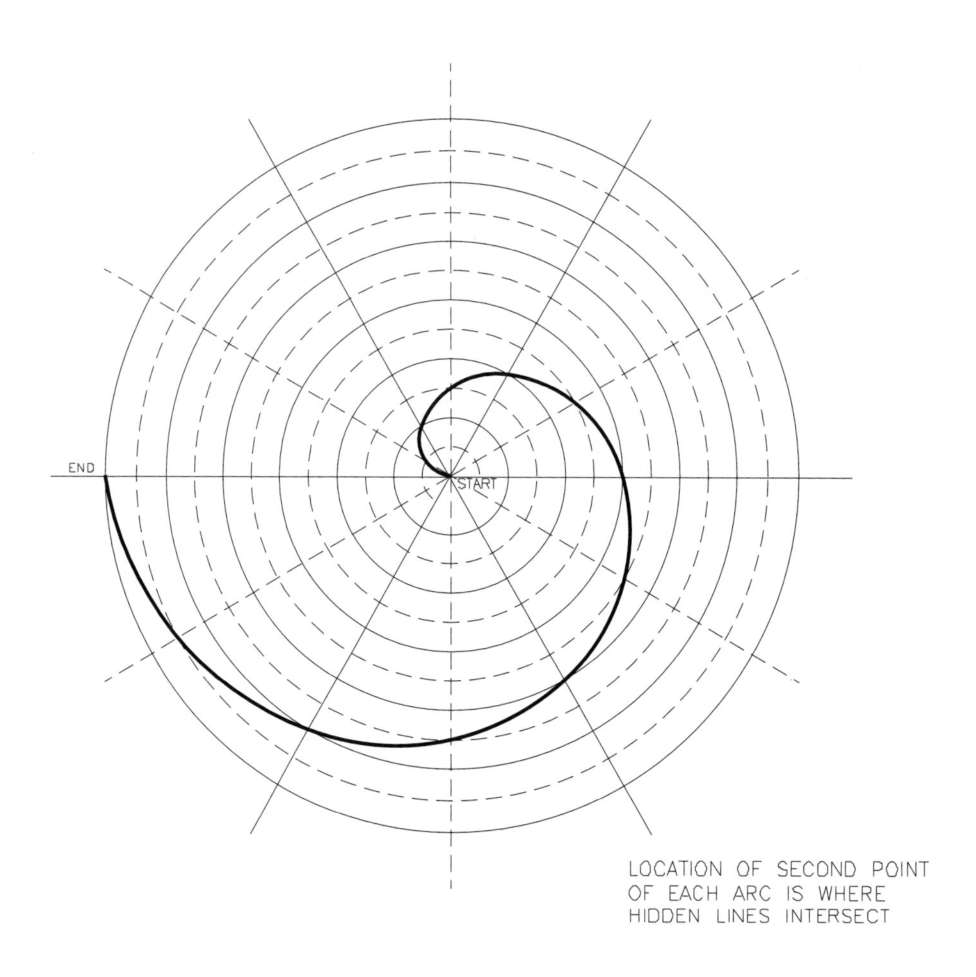

SOLID CIRCLE RADII	HIDDEN CIRCLE RADII
0.50	0.25
1.00	0.75
1.50	1.25
2.00	1.75
2.50	2.25
3.00	2.75

NOTE: THIS DRAWING IS USED
ON DRAWING 6–3

SAVE THIS DRAWING!

ARCHIMEDES SPIRAL

Drawing 6–1

DRAWING 6—2

Spiral Designs

These designs are different from other drawings in this book. There are no dimensions and you will use only edit commands now that the spiral is drawn. Below the designs is a list of the edit commands you will need. Don't be too concerned with precision. Some of your designs may come out slightly different from ours. When this happens, try to analyze the differences.

Drawing Suggestions

LIMITS = (0,0) (34,24)

These large limits will be necessary if you wish to draw all of these designs on the screen at once.

In some of the designs and in Drawing 6—3 you will need to rotate a copy of the spiral and keep the original in place. You can accomplish this using the grip edit rotate procedure with the copy option, or by making a copy of the objects before you enter the ROTATE command. Both procedures are listed following.

How to Rotate an Object and Retain the Original

Using Grip Edit

1. Select the spiral.
2. Pick the grip around which you want to rotate, or any of the grips if you are not going to use the grip as a base point for rotation.
3. Type "b" or select "Base point", if necessary. In this exercise the base point you choose for rotation will depend on the design you are trying to create.
4. Type "c" or select "Copy".
5. Show the rotation angle.
6. Press Enter to exit the grip edit system.

Using the ROTATE Command

1. Use COPY to make a copy of the objects you want to rotate directly on top of their originals. In other words, give the same point for the base point and the second point of displacement. When the copy is done, your screen will not look any different, but there will actually be two spirals there, one on top of the other.
2. Enter ROTATE and give "p" in response to the "Select objects:" prompt. This will select all the original objects from the last COPY sequence, without selecting the newly drawn copies.
3. Rotate as usual, choosing a base point dependent on the design you are creating.
4. After the ROTATE sequence is complete, you will need to do a REDRAW before the objects copied in the original position will be visible.

SPIRAL DESIGNS

(Make from Drawing 6-1)

Drawing 6-2

DRAWING 6—3

Grooved Hub

This drawing includes a typical application of the rotation technique just discussed. The hidden lines in the front view must be rotated 120 degrees and a copy retained in the original position. There are also good opportunities to use MIRROR, object snap, and TRIM.

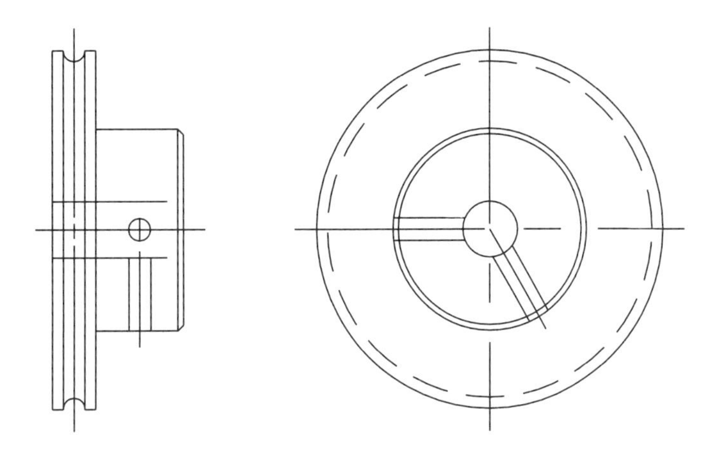

Drawing Suggestions

GRID = .5   LIMITS = (0,0) (18,12)
SNAP = .0625   LTSCALE = 1

- Draw the circles in the front view and use these to line up the horizontal lines in the left side view.
- There are several different planes of symmetry in the left side view, which suggests the use of mirroring. We leave it up to you to choose an efficient sequence.
- A quick method for drawing the horizontal hidden lines in the left side view is to use a quadrant osnap to begin a line at the top and bottom of the .62 diameter circle in the front view. Draw this line across to the back of the left side view, and use TRIM to erase the excess on both sides.
- The same method can be used to draw the two horizontal hidden lines in the front view. Snap to the top and bottom quadrants of the .25 diameter circle in the left side view as a guide and draw lines through the front view. Then trim to the 2.25 diameter circle and the .62 diameter circle.
- Once these hidden lines are drawn, rotate them, retaining a copy in the original position.

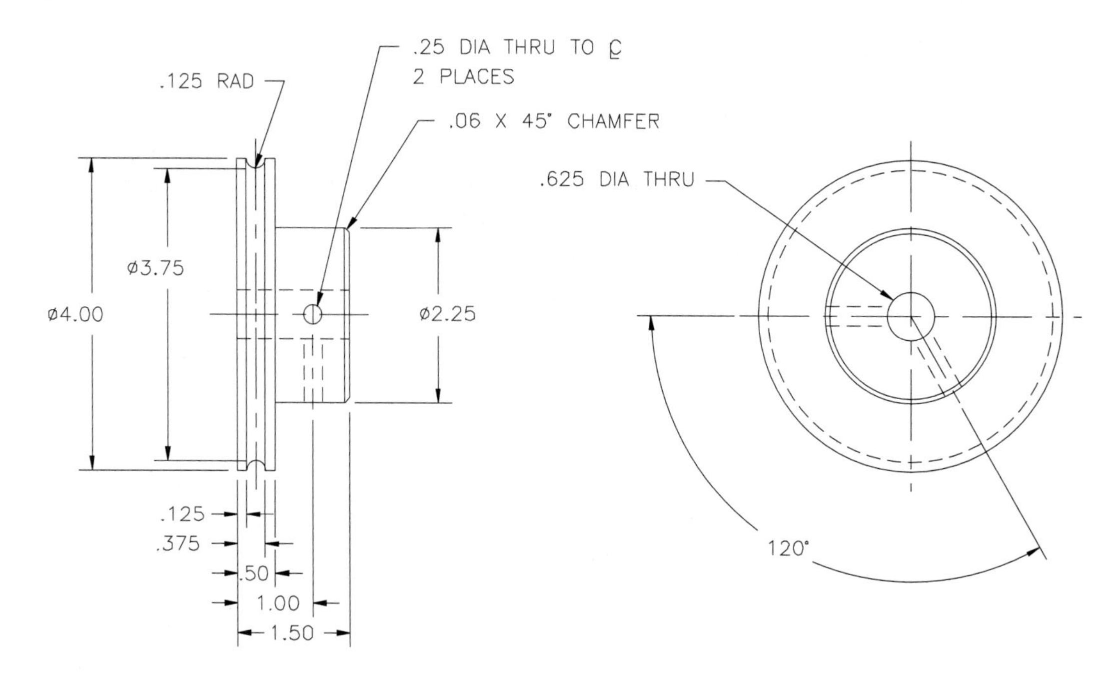

GROOVED HUB
Drawing 6–3

7

Text

OVERVIEW

Now it's time to add text to your drawings. In this chapter you will learn to find your way around Release 14's three text commands, TEXT, DTEXT, and MTEXT. In addition, you will learn many new editing commands that are often used with text but that are equally important for editing other objects.

SECTIONS

OBJECTIVES

After reading this chapter, you should be able to:

- Enter standard text using the DTEXT command.
- Enter standard text using different justification options.
- Use DDEDIT to edit previously drawn text.
- Check spelling of text in a drawing.
- Change fonts and styles.
- Change properties with MATCHPROP.
- Use SCALE to change the size of objects on the screen.
- Draw a title block.
- Draw gauges.
- Draw a control panel.

7.1 ENTERING LEFT-JUSTIFIED TEXT USING DTEXT

Release 14 provides three different commands for entering text in a drawing. TEXT, the oldest of the three, allows you to enter single lines of text at the command prompt. DTEXT uses all of the same options as TEXT but shows you the text on the screen as you enter it, as well as allowing you to enter multiple lines of text and to backspace through them to make corrections. MTEXT allows you to type paragraphs of text in a dialog box and then positions them in a windowed area in your drawing. All of these commands have numerous options for placing text and a variety of fonts to use and styles that can be created from them.

We will focus on the DTEXT command. Most of what you learn will apply to the TEXT command as well, but DTEXT has more features and is easier to use.

To prepare for this exercise, open a new drawing using the B template and draw a 4.00 horizontal line beginning at (1,1). Then create a 6 row by 1 column array with 2.00 between rows, as shown in *Figure 7.1*.

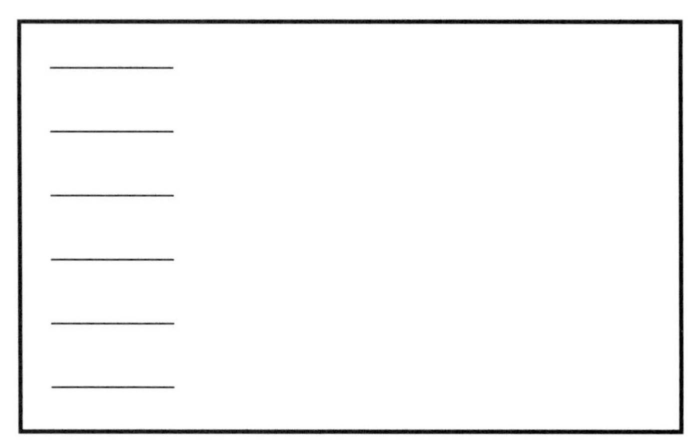

Figure 7.1.

These lines are for orientation in this exercise only; they are not essential for drawing text.

⌖ **Type "dt" or open the Draw menu, highlight Text and then select Single line text, as shown in *Figure 7.2***

There is a text tool on the Draw toolbar, but it enters the MTEXT command.

Either of these methods will enter the DTEXT command and you will see a prompt with three options in the command area:

Justify/Style/<Start point>:

Figure 7.2.

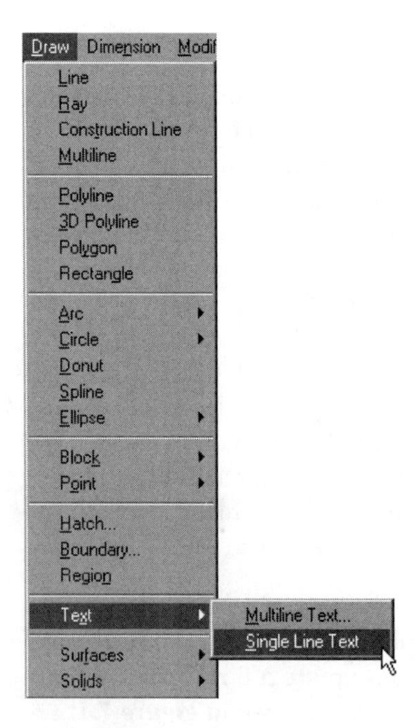

"Style" will be explored in Task 7.8. In this task we will be looking at different options for placing text in a drawing. These are all considered text justification methods and will be listed if you choose the "Justify" option at the command prompt.

First we will use the default method by picking a start point. This will give us left-justified text, inserted left to right from the point we pick.

⌖ **Pick a start point at the left end of the upper line.**

Look at the prompt that follows and be sure that you do not attempt to enter text yet:

Height <0.20>:

This gives you the opportunity to set the text height. The number you type specifies the height of uppercase letters in the units you have specified for the current drawing. For now we will accept the default height.

⌖ **Press Enter to accept the default height (0.20).**

The prompt that follows allows you to place text in a rotated position.

Rotation angle <0>:

The default of 0 degrees orients text in the usual horizontal manner. Other angles can be specified by typing a degree number relative to the polar coordinate system, or by showing a point. If you show a point, it will be taken as the second point of a baseline along which the text string will be placed. For now we will stick to horizontal text.

⌖ **Press Enter to accept the default angle (0).**

Now, at last, it is time to enter the text itself. AutoCAD prompts:

Text:

Notice also that a small text cursor has appeared at the start point on your screen. Move your cross hairs away from this point and you will see the cursor clearly. This shows where the first letter you type will be placed. It is a feature of DTEXT that you would not find in the TEXT command. For our text we will type the word "Left" since this is an example of left-justified text. Watch the screen as you type and you will see dynamic text at work.

⌖ **Type "Left" and press Enter.**

(Remember, you cannot use the space bar in place of the enter key when entering text.)

Notice that the text cursor jumps down below the line when you hit Enter. Also notice that you are given a second "Text:" prompt in the command area.

⌖ **Type "Justified" and press Enter.**

The text cursor jumps down again and another "Text:" prompt appears. This is how DTEXT allows for easy entry of multiple lines of text directly on the screen in a drawing. To exit the command, you need to press Enter at the prompt.

⌖ **Press Enter to exit DTEXT.**

This completes the process and returns you to the command prompt.

Figure 7.3 shows the left-justified text you have just drawn, along with the other options as we will demonstrate in the next section.

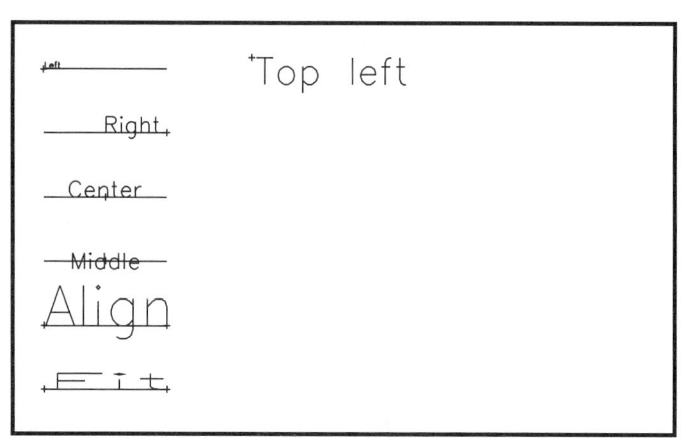

Figure 7.3.

7.2 USING OTHER TEXT JUSTIFICATION OPTIONS

We will now proceed to try out some of the other text placement options, beginning with right-justified text, as shown on the second line of *Figure 7.3.* We will also specify a change in height. The remaining options are shown in *Figure 7.3* so that you can try them out on your own.

Right-Justified Text

Right-justified text is constructed from an end point backing up, right to left.

⌗ **Repeat the DTEXT command.**

You will see the same prompt as before.

⌗ **Type "r" for right justified text.**

Now AutoCAD prompts you for an end point instead of a start point:

End point:

We will choose the right end of the second line.

⌗ **Point to the right end of the second line.**

This time we will change the height to .50. Notice that AutoCAD gives you a rubber band from the end point. It can be used to specify height and rotation angle by pointing, if you like.

⌗ **Type ".5" or show a height of .50 by pointing.**

⌗ **Press Enter to retain 0 degrees of rotation.**

You are now prompted to enter text.

⌗ **Type "Right" and press Enter.**

Here you should notice an important fact about the DTEXT command. Initially, DTEXT ignores your justification choice, entering the letters you type from left to right as usual. The justification will be carried out only when you exit the command.

At this point you should have the word "Right" showing to the right of the second line. Watch what happens when you press Enter a second time to exit the command.

⌗ **Press Enter.**

The text will jump to the left, onto the line. Your screen should now include the second line of text in right-justified position as shown in *Figure 7.3.*

Centered Text

Centered text is justified from the bottom center of the text.

⌖ **Repeat the DTEXT command.**
⌖ **Type "c".**
AutoCAD prompts:

Center point:

⌖ **Point to the midpoint of the third line.**
⌖ **Press Enter to retain the current height, which is now set to .50.**
⌖ **Press Enter to retain 0 degrees of rotation.**
⌖ **Type "Center" and press Enter.**

Notice again how the letters are displayed on the screen in the usual left to right manner.

⌖ **Press Enter again to complete the command.**

The word "Center" should now be centered as shown in *Figure 7.3.*

Middle, Aligned, and Fit Text

Looking at the other three lines of Figure 7.3 you will see middle text, aligned text, and text stretched to fit on the line. You can try these options using the procedures just described. Repeat Dtext, type in an option, and pick a point or points on the line.

Middle text (type "m") is justified from the middle of the text horizontally and vertically, instead of from the bottom.

Aligned text (type "a") is placed between two specified points. The height of the text is calculated proportional to the distance between the two points and the text is drawn along the line between the two points.

The Fit option (type "f") is similar to the Aligned option, except that the specified text height is retained. Text will be stretched horizontally to fill the line without a change in height.

Other Justification Options

There are additional justification options that are labeled with various combinations of top, middle, bottom, left, and right. The letter options are shown in Figure 7.4. As shown on the chart and the screen menu, T is for top, M is for middle, and B is for bottom. L, C, and R stand for left, center, and right. All of these options work the same way. Just enter TEXT or DTEXT and then the one or two letters of the option.

7.3 EDITING TEXT WITH DDEDIT

There are several ways to modify text that already exists in your drawing. You can change wording and spelling as well as properties such as layer, style, and justification. Commands that you will use to alter text include CHANGE, CHPROP, DDCHPROP, DDEDIT, and DDMODIFY. CHANGE, DDMODIFY, and DDEDIT can be used to change words as well as text properties, CHPROP AND DDCHPROP can change only properties.

For text editing you will most often use DDEDIT and DDMODIFY. We will begin with some simple DDEDIT text editing.

⌖ **Type "ddedit" or open the Modify menu, highlight Object and select Text.**

TEXT JUSTIFICATION	
<START POINT> TYPE ABBREVIATION	TEXT POSITION + INDICATES START POINT or PICK POINT
A	ALIGN
F	FIT
C	CENTER
M	MIDDLE
R	RIGHT
TL	TOP LEFT
TC	TOP CENTER
TR	TOP RIGHT
ML	MIDDLE LEFT
MC	MIDDLE CENTER
MR	MIDDLE RIGHT
BL	BOTTOM LEFT
BC	BOTTOM CENTER
BR	BOTTOM RIGHT

Figure 7.4.

There is also an Edit Text tool that you can access by opening the Modify II toolbar. Regardless of how you enter the command, AutoCAD will prompt you to select an object for editing:

<Select an annotation object>/Undo:

Annotation objects will include all objects that have text, including text, dimensions, and attributes. Dimensions are the subject of Chapter 8.

We will select one line of the angled text.

⌖ Select the words "These lines" by clicking on any of the letters.

As soon as you select the text, it will appear in a small edit box, as illustrated in *Figure 7.5.* This is the DDEDIT dialog box for text created with DTEXT and TEXT. Now add the word "three" to the middle of this line.

⌖ Move the dialog box arrow to the center of the text in the edit box, between "These" and "lines", and press the pick button.

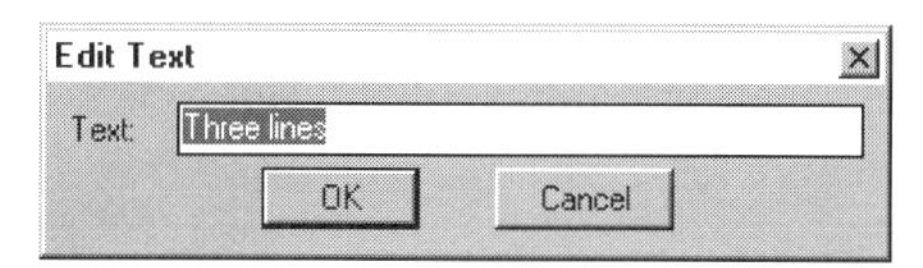

Figure 7.5.

The text should no longer be highlighted and a flashing cursor should be present indicating where text will be added if you begin typing.

⌖ Type the word "three" and add a space so that the line reads "These three lines".

⌖ Click on "OK".

The dialog box will disappear and your text will be redrawn with the word "three" added as shown in *Figure 7.6.*

7.4 USING THE SPELL COMMAND

AutoCAD's SPELL command is simple to use and will be very familiar to anyone who has used spell checkers in word processing programs. We will use SPELL to check the spelling of all the text we have drawn so far.

⌖ Type "sp", select Spelling from the Tools menu, or select the Spelling tool from the Standard toolbar, as shown in *Figure 7.7.*

You will see a "Select objects:" prompt on the command line. At this point you could point to individual objects. Any object may be selected, although no checking will be done if you select a line, for example.

For our purposes we will use an "All" option to check all the spelling in the drawing.

⌖ Type "all".

AutoCAD will continue to prompt for object selection until you press Enter.

⌖ Press Enter to end object selection.

This will bring you to the Check Spelling dialog box shown in *Figure 7.8.* If you have followed the exercise so far and not misspelled any words along the way, you will see "Mtext" in the Current word box and "Text" as a suggested correction. We will

Figure 7.7.

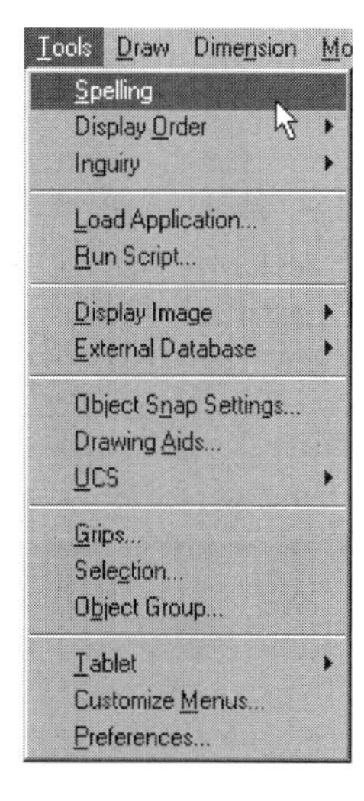

Figure 7.6.

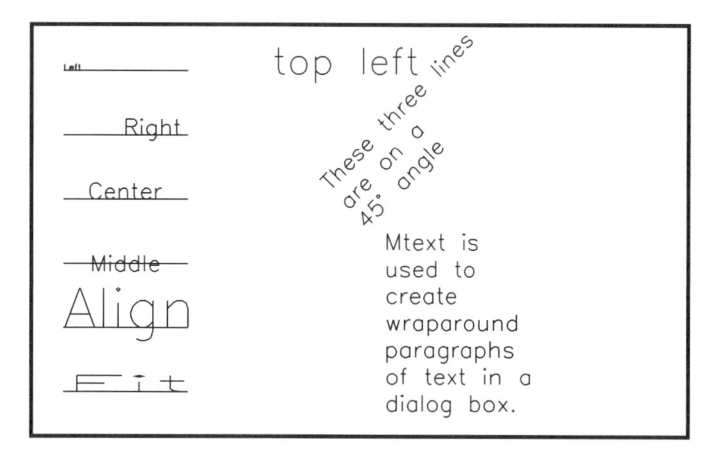

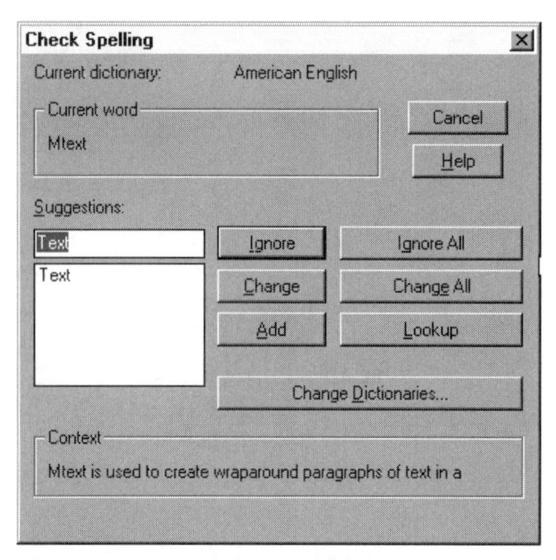

Figure 7.8.

ignore this change, but before you leave SPELL, look at what is available: You can ignore a word the checker does not recognize, or change it. You can change a single instance of a word, or change all instances in the currently selected text. You can add a word to a customized dictionary or you can change to another dictionary.

Click on "Ignore".

If your drawing does not contain other spelling irregularities, you should now see an AutoCAD message that says:

Spelling check complete.

Click on "OK" to end the spell check.

If you have made any corrections in spelling they will be incorporated into your drawing at this point.

7.5 CHANGING FONTS AND STYLES

By default, the current text style in any AutoCAD drawing is one called "STANDARD". It is a specific form of a font called "txt" that comes with the software. All the text you have entered so far has been drawn with the standard style of the "txt" font.

Changing fonts is a simple matter. However, there is room for confusion in the use of the words "style" and "font." You can avoid this confusion if you remember that fonts are the basic patterns of character and symbol shapes that can be used with the TEXT, DTEXT, and MTEXT commands, while styles are variations in the size, orientation, and spacing of the characters in those fonts. It is possible to create your own fonts, but for most of us this is an esoteric activity. In contrast, creating your own styles is easy and practical.

We will begin by creating a variation of the STANDARD style you have been using.

Type "st" or select "Text Style" from the Format menu.

Either method will open the Text Style dialog box shown in *Figure 7.9*. You will probably see STANDARD listed in the Style name box. However, if anyone has used text commands in your prototype drawing, it is possible that there will be other styles listed. In this case you should open the list with the arrow on the right and select STANDARD. It is certain to be there, because it is created automatically. We will create

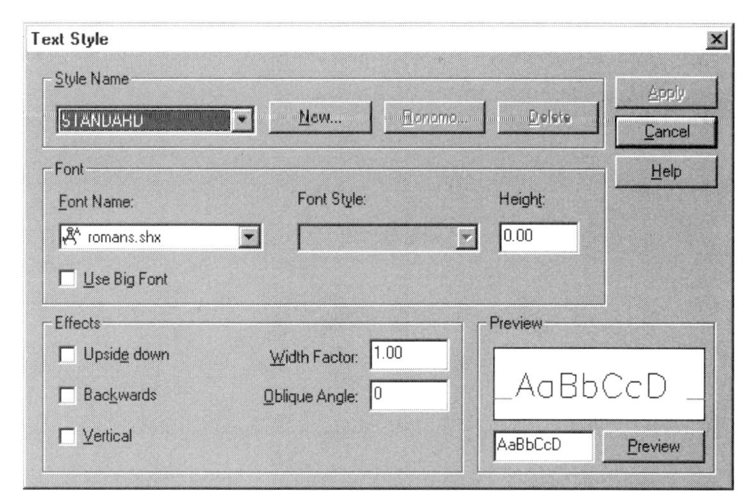

Figure 7.9.

our own variation of the STANDARD style and call it "VERTICAL". It will use the same "txt" character font, but will be drawn down the display instead of across.

Click in the "Vertical" check box.

We will also give this style a fixed height and a width factor. Notice that the current height is 0.00. This does not mean that your characters will be drawn 0.00 units high. It means that there will be no fixed height, so you will be able to specify a height whenever you use this style. STANDARD currently has no fixed height, so VERTICAL has inherited this setting. Try giving our new "VERTICAL" style a fixed height.

Double click in the Height edit box and then type ".5".

Double click in the Width Factor box and type "2".

Click on "New".

AutoCAD opens a smaller dialog box that asks for a name for the new text style.

Type "vertical".

Click on OK.

This brings you back to the Text Style dialog box, with the new style listed in the style name area.

Click on Close to exit the Text Style dialog.

The new VERTICAL style is now current. To see it in action you will need to enter some text

Type "dt" or open the Draw menu, highlight Text and then Single Line text.

Pick a start point, as shown by the blip near the letter "V" in *Figure 7.10.*

Notice that you are not prompted for a height because the current style has height fixed at .50.

Press Enter to retain 270 degrees of rotation.

Type "Vertical".

Press Enter to end the line.

Before going on, notice that the DTEXT text placement box has moved up to begin a new column of text next to the word vertical.

Press Enter to exit DTEXT.

Your screen should resemble *Figure 7.10.*

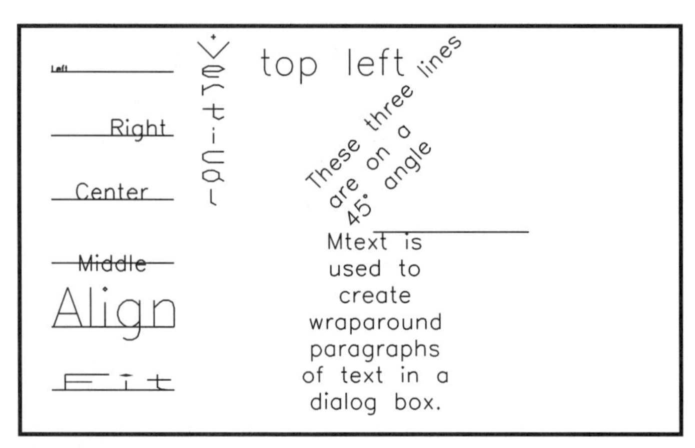

Figure 7.10.

Switching the Current Style

There are now two styles defined in your drawing. All new text is created in the current style. The style of previously drawn text can be changed, as we will show later. Once you have a number of styles defined in a drawing, you can switch from one to another with the Style option of the TEXT and DTEXT commands or by Selecting a text style from the Text Style dialog box.

NOTE: If you change the definition of a text style, all text previously drawn in that style will be regenerated with the new style specifications.

7.6 CHANGING PROPERTIES WITH MATCHPROP

MATCHPROP is a very efficient command, new with Release 14, that lets you match all or some of the properties of an object to those of another object. Properties that can be transferred from one object to another, or to many others, include layer, linetype, color, lintype scale. These settings are common to all AutoCAD entities. Other properties that only relate to specific types of entities are thickness, text style, dimension style, and hatch style. In all cases the procedure is the same.

Here we will use MATCHPROP to change some previously drawn text to the new "Third" style.

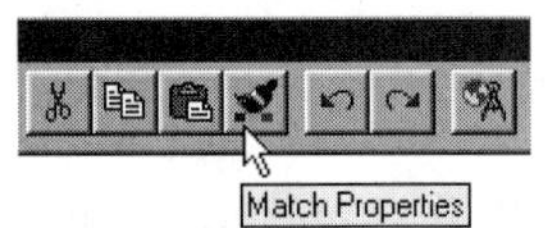

Figure 7.11.

Select the Match Properties tool from the Standard toolbar, as shown in *Figure 7.11.*

AutoCAD will prompt:

Select Source Object:

You can have many destination objects but only one source object.

Select the text "Roman Duplex", drawn in the last task.

AutoCAD prompts:

Settings/<Select Destination Objec(s)>:

At this point you can limit the settings you want to match by typing "s" for settings, or you can select destination objects, in which case all properties will be matched.

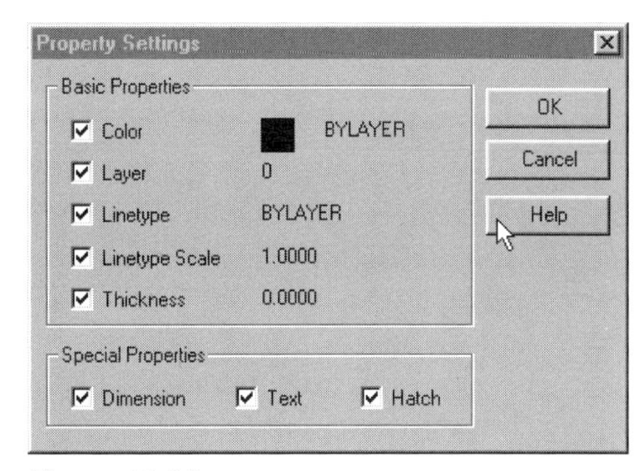

Figure 7.12.

⌖ Type "s".

This opens the Property Settings dialog box, shown in *Figure 7.12*. The Basic properties area on the top shows basic properties that can be changed and the settings that will be used based on the source object you have selected.

At the bottom you will see Dimension, Text, and Hatch in the Special Properties area. These refer to styles which have been defined in your drawing for these types of objects. If any of these is deselected, then match properties will ignore styles and will only match the basic properties checked.

⌖ Click on OK to exit the dialog.

AutoCAD returns to the screen with the same prompt as before.

⌖ Select the words "Align" and "Vertical".

These two word will be redrawn in the Third style, as shown in *Figure 7.13*.

The CHANGE command

The CHANGE command is an older AutoCAD command that has been largely replaced by other editing commands. CHANGE works at the command line and allows you to modify all of the basic text properties, including insertion point, style, height, rotation angle, and the text itself. You can select as many objects as you like and CHANGE will cycle through them. In addition, CHANGE can be used to alter the end points of lines and the size of circles. With lines, CHANGE will perform a function similar to the

Figure 7.13.

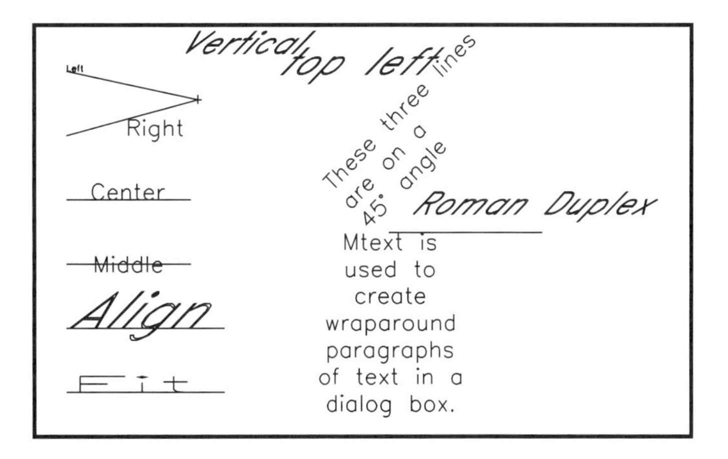

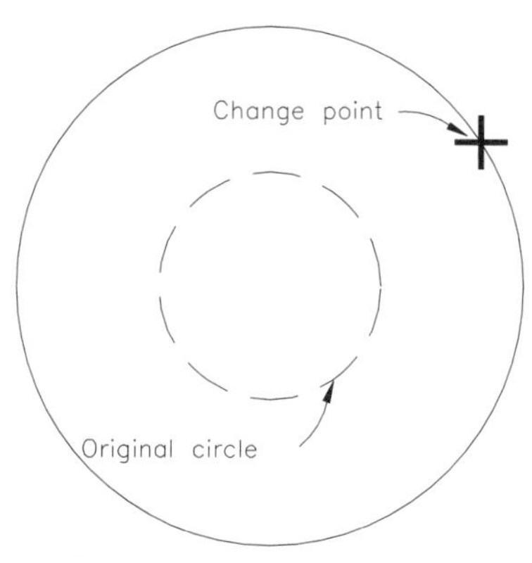

Figure 7.14.

EXTEND command, but without the necessity of defining an extension boundary. With circles, CHANGE will cause them to be redrawn so that they pass through a designated "change point." The effects of change points on lines and circles are shown in *Figure 7.14.* Notice in particular the way lines will react to a change point with ortho on.

7.7 SCALING PREVIOUSLY DRAWN ENTITIES

Any object or group of objects can be scaled up or down using the SCALE command or the grip edit scale mode. In this exercise we will practice scaling some of the text and lines that you have drawn on your screen. Remember, however, that there is no special relationship between SCALE and text and that other types of entities can be scaled just as easily.

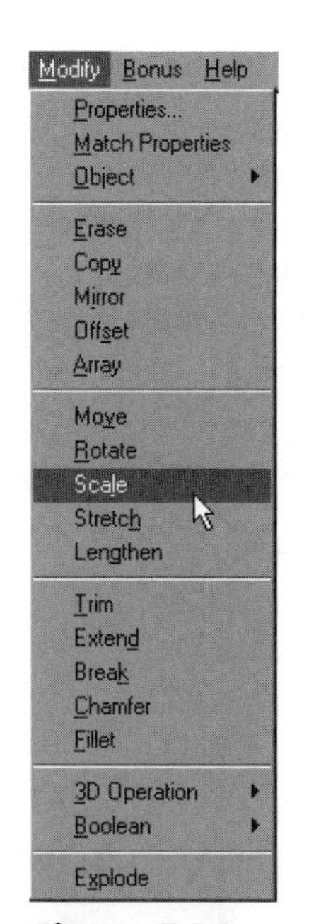

Figure 7.15.

⌖ **Type "Sc", select Scale from the Modify menu, or the Scale tool from Modify toolbar, as shown in *Figure 7.15.***

AutoCAD will prompt you to select objects.

⌖ **Use a crossing box (right to left) to select the set of six lines and text drawn in Task 1.**

⌖ **Press Enter to end selection.**

You will be prompted to pick a base point:

Base point:

The concept of a base point in scaling is critical. Imagine for a moment that you are looking at a square and you want to shrink it using a scale-down procedure. All the sides will, of course, be shrunk the same amount, but how do you want this to happen? Should the lower left corner stay in place and the whole square shrink toward it? Or should everything shrink toward the center? Or toward some other point on or off the square (see *Figure 7.16).*This is what you will tell AutoCAD when you pick a base point.

⌖ **Pick a base point at the left end of the bottom line of the selected set (the blip in *Figure 7.17*).**

AutoCAD now needs to know how much to shrink or enlarge the objects you have selected:

<Scale factor>/Reference:

We will get to the reference method in a moment. When you enter a scale factor, all lengths, heights, and diameters in your set will be multiplied by that factor and

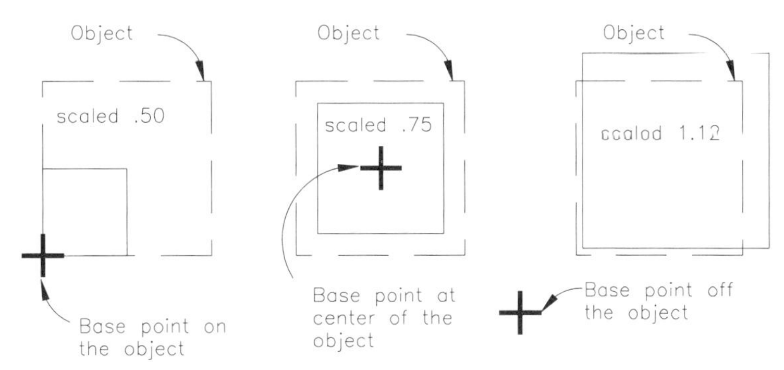

Figure 7.16.

redrawn accordingly. Scale factors are based on a unit of 1. If you enter .5, objects will be reduced to half their original size. If you enter 2, objects will become twice as large.

⌗ **Type ".5" and press Enter.**

Your screen should now resemble *Figure 7.17.*

SCALEing by Reference

This option can save you from doing the arithmetic to figure out scale factors. It is useful when you have a given length and you know how large you want that length to become after the scaling is done. For example, we know that two of the lines we just scaled are now 2.00 long (the middle two that we did not alter with CHANGE). Let's say that we want to scale them again to become 2.33 long (a scale factor of 1.165, but who wants to stop and figure that out?). This could be done using the following procedure:

1. Enter the SCALE command.
2. Select the "previous" set.
3. Pick a base point.
4. Type "r" or select "reference".
5. Type "2" for the reference length.
6. Type "2.33" for the new length.

Figure 7.17.

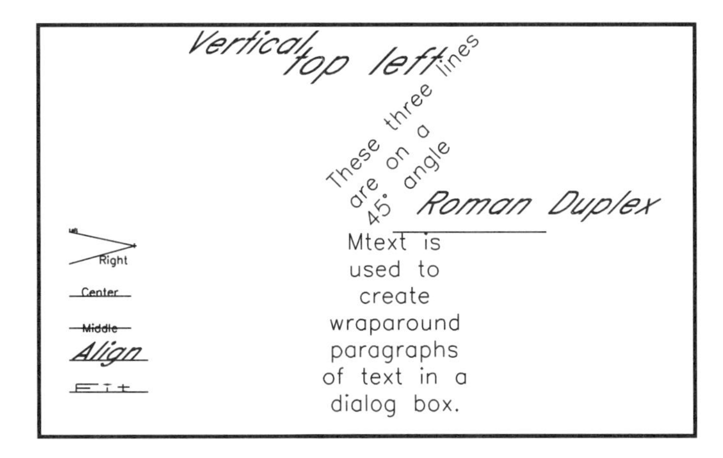

NOTE: You can also perform reference scaling by pointing. In the procedure above, you could point to the ends of the 2.00 line for the reference length and then show a 2.33 line for the new length.

Scaling with Grips

Scaling with grips is very similar to scaling with the SCALE command. To illustrate this, try using grips to return the text you just scaled back to its original size.

⌗ **Use a window or crossing box to select the six lines and the text drawn in Task 1 again.**

There will be a large number of grips on the screen, three on each line and two on most of the text entities. Some of these will overlap or duplicate each other.

⌗ **Pick the grip at the lower left corner of the word "Fit," the same point used as a base point in the last scaling procedure.**

⌗ **Press the enter button on your cursor to open the cursor menu and then select "Scale", or press Enter three times to bypass stretch, move, and rotate at the command prompt.**

⌗ **Move the cursor slowly and observe the dragged image.**

AutoCAD will use the selected grip point as the base point for scaling unless you specify that you want to pick a different base point.

Notice that you also have a reference option as in the SCALE command. Unlike the SCALE command you also have an option to make copies of your objects at different scales.

As in SCALE, the default method is to specify a scale factor by pointing or typing.

⌗ **Type "2" or show a length of 2.00.**

Your text will return to its original size and your screen will resemble *Figure 7.24* again.

⌗ **Press Esc twice to clear grips.**

7.8 REVIEW MATERIAL

Questions

1. You have drawn two lines of text using DTEXT and have left the command to do some editing. You discover that a third line of text should have been entered with the first two lines. What procedure will allow you to efficiently add the third line of text so that it is spaced and aligned with the first two, as if you had never left DTEXT?
2. What is the difference between center justified text and middle justified text?
3. What aspect of text can be changed with DDEDIT? What is the purpose of MATCHPROP?
4. How do you check all the spelling in your drawing at once?
5. What is the difference between a font and a style?
6. What can happen if you choose the wrong base point when using the SCALE command?
7. How would you use SCALE to change a 3.00 line to 2.75?

COMMANDS

Draw	Modify	Data	Object Properties
CHANGE	STYLE		DDEMODES
	CHPROP		MATCHPROP
	DDCHPROP		
	DDEDIT	Tools	
	DDMODIFY	SPELL	
	SCALE		

Drawing Problems

1. Draw a 6 × 6 square. Draw the word "Top", on top of the square, 0.4 units high, centered on the midpoint of the top side of the square.
2. Draw the word "Left", 0.4 units high, centered on the left side of the square.
3. Draw the word "Right", 0.4 units high, centered on the right side of the square.
4. Draw the word "Bottom", 0.4 units high, below the square so that the top of the text is centered on the midpoint of the bottom side of the square.
5. Draw the words "This is the middle", inside the square, 0.4 units high, so that the complete text wraps around within a 2 unit width and is centered on the center point of square.

DRAWING 7—1

Title Block

This title block will give you practice in using a variety of text styles and sizes. You may want to save it and use it as a title block for future drawings. In Chapter 10 we will show you how to insert one drawing into another, so you will be able to incorporate this title block into any drawing.

QTY REQ'D
DESCRIPTION
PART NO.
ITEM NO.
BILL OF MATERIALS
UNLESS OTHERWISE SPECIFIED DIMENSIONS ARE IN INCHES
DRAWN BY: B. A. Designer
DATE
CSA INC.
REMOVE ALL BURRS & BREAK SHARP EDGES
APPROVED BY:
TOLERANCES
FRACTIONS ± 1/64
ANGLES ± 0°-15'
DECIMALS
XX ± .01
XXX ± .005
ISSUED:
DRAWING TITLE:
MATERIAL:
FINISH:
SIZE C
CODE IDENT NO. 38178
DRAWING NO.
REV.
SCALE:
DATE:
SHEET OF

Drawing Suggestions

GRID = 1
SNAP = .0625

- Make ample use of TRIM as you draw the line patterns of the title block. Take your time and make sure that at least the major divisions are in place before you start entering text into the boxes.
- Set to the "text" layer before entering text.
- Use DTEXT with all the STANDARD, .09, left-justified text. This will allow you to do all of these in one command sequence, moving the cursor from one box to the next and entering the text as you go.
- Remember that once you have defined a style, you can make it current using the TEXT or DTEXT commands. This will save you from having to restyle more than necessary.
- Use "%%D" for the degree symbol and "%%P" for the plus or minus symbol.

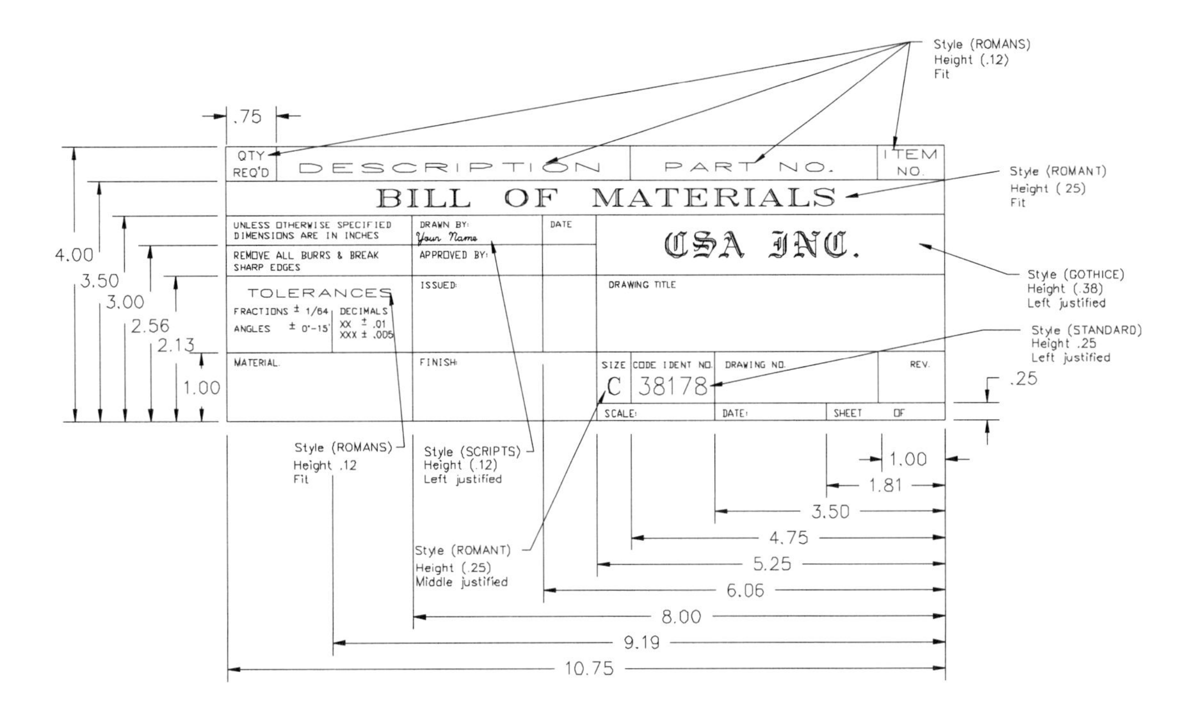

TITLE BLOCK

Drawing 7–1

DRAWING 7—2

Gauges

This drawing will teach you some typical uses of the SCALE and CHANGE commands. Some of the techniques used will not be obvious, so read the suggestions carefully.

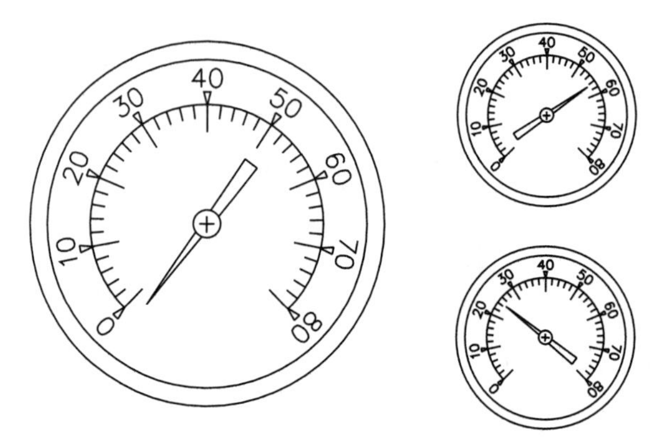

Drawing Suggestions

GRID = .5
SNAP = .125

- Draw three concentric circles at diameters of 5.0, 4.5, and 3.0. The bottom of the 3.0 circle can be trimmed later.
- Zoom in to draw the arrow-shaped tick at the top of the 3.0 circle. Then draw the .50 vertical line directly below it and the number "0" (middle-justified text) above it.
- These three objects can be arrayed to the left and right around the perimeter of the 3.0 circle using angles of +135 and −135 as shown.
- Use the CHANGE command to change the arrayed zeros into 10, 20, 30, etc.
- Draw the .25 vertical tick directly on top of the .50 mark at top center and array it left and right. There will be 20 marks each way.
- Draw the needle horizontally across the middle of the dial.
- Make two copies of the dial; use SCALE to scale them down as shown. Then move them into their correct positions.
- Rotate all three needles as shown.

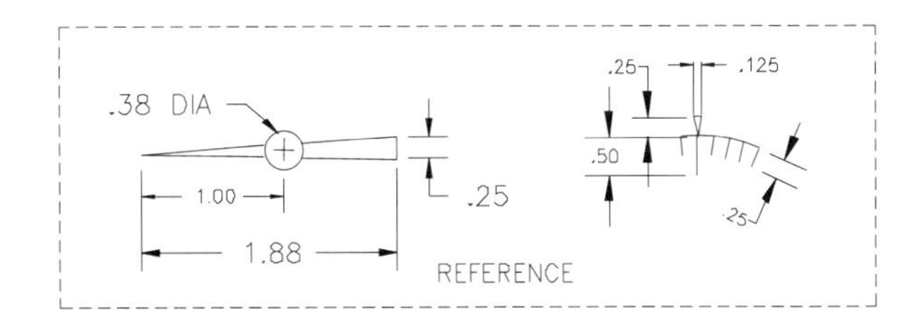

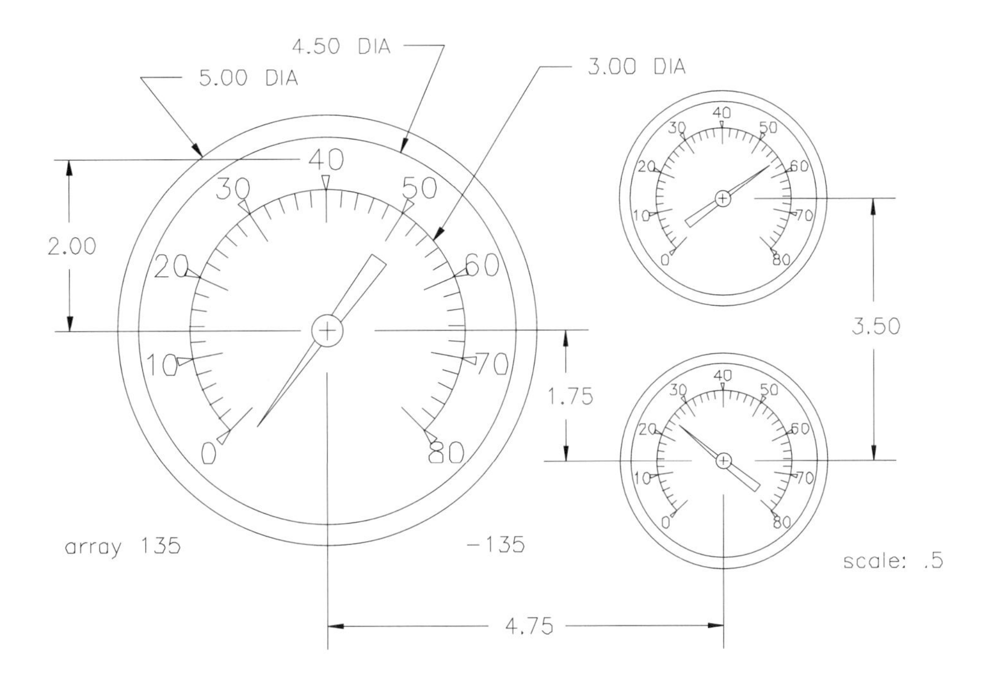

GAUGES
Drawing 7-2

DRAWING 7—3

Control Panel

Done correctly, this drawing will give you a good feel for the power of the commands you now have available to you. Be sure to take advantage of combinations of ARRAY and CHANGE as described. Also, read the suggestion on moving the origin before you begin.

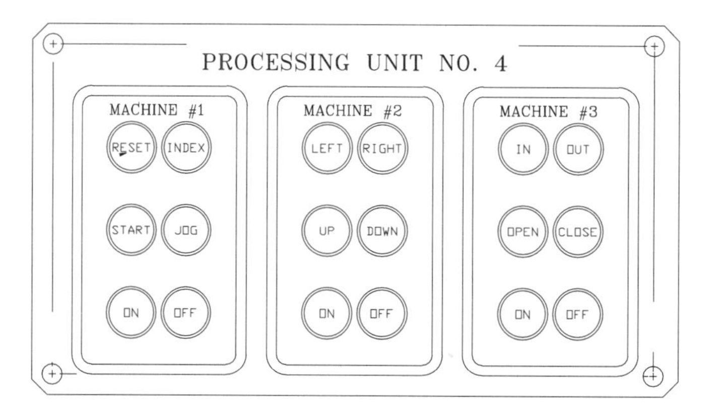

Drawing Suggestions

GRID = .50
SNAP = .0625

- After drawing the outer rectangles, draw the double outline of the left button box, and fillet the corners. Notice the different fillet radii.
- Draw the "on" button with its text at the bottom left of the box. Then array it 2 × 3 for the other buttons in the box.
- CHANGE the lower right button text to "off" and draw the MACHINE # text at the top of the box.
- ARRAY the box 1 × 3 to create the other boxes.
- CHANGE text for buttons and machine numbers as shown.

Moving the Origin with the Ucs Command The dimensions of this drawing are shown in ordinate form, measured from a single point of origin in the lower left-hand corner. In effect, this establishes a new coordinate origin. If we move our origin to match this point, then we will be able to read dimension values directly from the coordinate display. This may be done by setting the lower left-hand limits to (–1,–1). But it also may be done using the UCS command to establish a User Coordinate System with the origin at a point you specify. User Coordinate Systems are discussed in depth in Chapter 12. For now, here is a simple procedure:

1. Type "ucs".
2. Type "o" for the "Origin" option.
3. Point to the new origin.

That's all there is to it. Move your cursor to the new origin and watch the coordinate display. It should show "0.00,0.00,0.00", and all values will be measured from there.

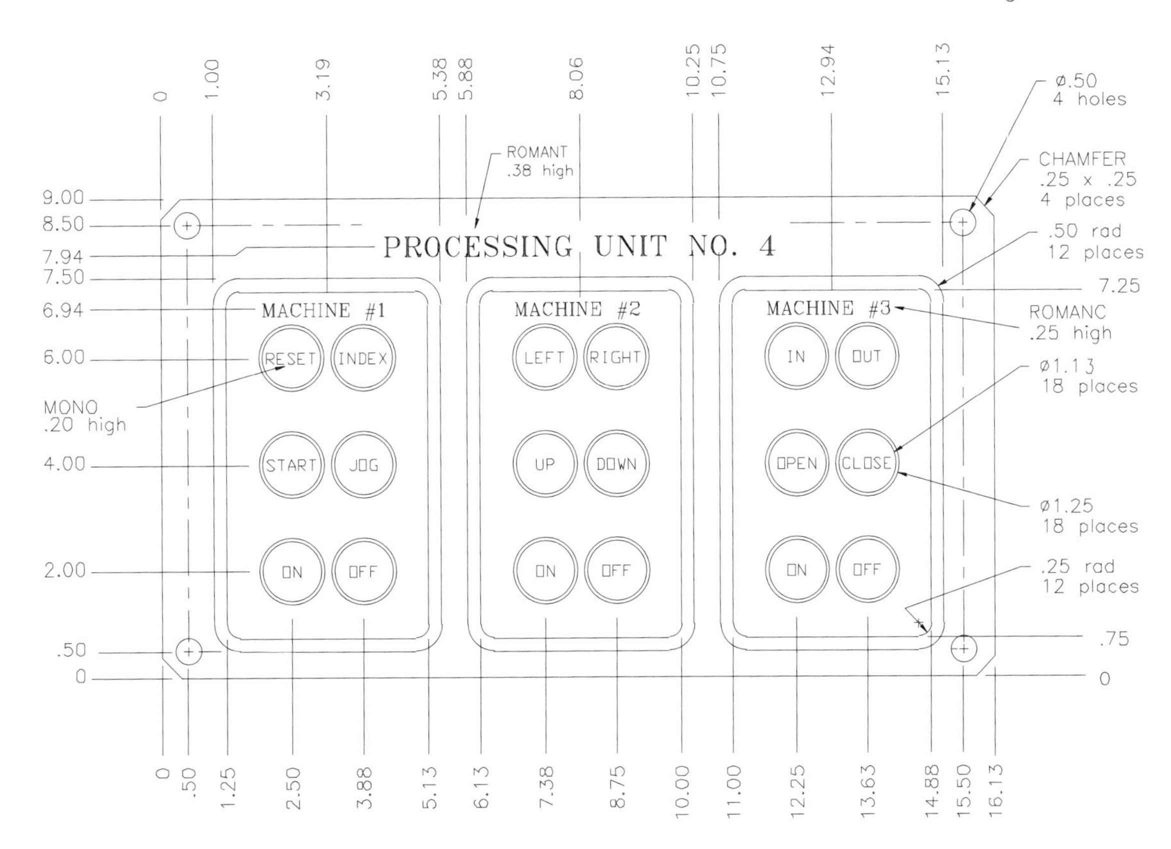

CONTROL PANEL

Drawing 7-3

8

Dimensions

OVERVIEW

The ability to dimension your drawings and add crosshatch patterns will greatly enhance the professional appearance and utility of your work. AutoCAD's dimensioning feature is a complex system of commands, subcommands, and variables that automatically measure objects and draw dimension text and extension lines. With AutoCAD's dimensioning tools and variables, you can create dimensions in a wide variety of formats, and these formats can be saved as styles. The time saved through not having to draw each dimension line by line is among the most significant advantages of CAD.

SECTIONS

OBJECTIVES

After reading this chapter, you should be able to:

- Define and save a dimension style.
- Draw linear dimensions (horizontal, vertical, and aligned).
- Draw baseline and continued linear dimensions.
- Draw angular dimensions.
- Draw center marks and diameter and radius dimensions.
- Add cross-hatching to previously drawn objects.
- Draw a flanged wheel.
- Draw a shower head.
- Draw a plot plan.

8.1 CREATING AND SAVING A DIMENSION STYLE

Dimensioning in AutoCAD is highly automated and very easy compared to manual dimensioning. In order to achieve a high degree of automation while still allowing for the broad range of flexibility required to cover all dimension styles, the AutoCAD dimensioning system is necessarily quite complex. In the exercises that follow we will guide you through the system, show you some of what is available, and give you a good foundation for understanding how to get what you want out of AutoCAD dimensioning. We will create a basic dimension style and use it to draw standard dimensions and tolerances. We will leave it to you to explore the many variations that are possible.

In Release 14 it is best to begin by defining a dimension style. A dimension style is a set of dimension variable settings that control the text and geometry of all types of AutoCAD dimensions. Prior to Release 13, the default styles for most dimensions would use the drawing units set through the Units Control dialog box or the UNITS command. Release 13 and 14 dimensions, however, dimensioning units must be set separately. We recommend that you create the new dimension style in your template drawing and save it. Then you will not have to make these changes again when you start new drawings.

⌗ **To begin this exercise, open the B template drawing.**

We will make changes in dimension style settings in the template drawing so that all dimensions showing distances will be presented with two decimal places and angular dimensions will have no decimals. This will become the default dimension setting in any drawing created using the B template.

⌗ **Type "ddim", select Dimension style from the Format menu, Style from the Dimension menu, or open the Dimension toolbar and select the Dimension tool (see *Figure 8.1*).**

TIP: Dimensioning is a good example of a case where you may want to open a toolbar and leave it open for awhile. Since dimensioning is often saved for last in a drawing and may be done all at once, bringing up the toolbar and keeping it on your screen while you dimension may be most efficient.

Any of the above methods will call the Dimension Styles dialog box shown in *Figure 8.2*. You will see that the current dimension style is called "Standard". Among other things, the AutoCAD Standard dimension style uses four-place decimals in all dimensions, including angular dimensions.

On the left below the Dimension Style name box you will see the Family box with a set of radio buttons for the different types of dimensions. A family is a complete group of dimension settings for the different types of dimensions. The types of dimensions are shown with radio buttons: Linear, Radial, Angular, etc. You can set these individually or use the Parent radio button to set them all at once to a single format. In this exercise we will first use the Parent button to set all types of dimensions to two-place decimals. Then we will set angular dimensions individually to show no decimal places.

But first, let's give the new dimension style a name. We will use "B" to go along with our B prototype.

⌗ **Double click in the Name: edit box to highlight the word "Standard".**

⌗ **Type "B".**

⌗ **Click on "Save".**

This will create the new dimension style and make it current, so that subsequent dimensions will be drawn using the B style.

⌗ **Check to make sure that the Parent radio button is selected. It should be by default.**

Figure 8.1.

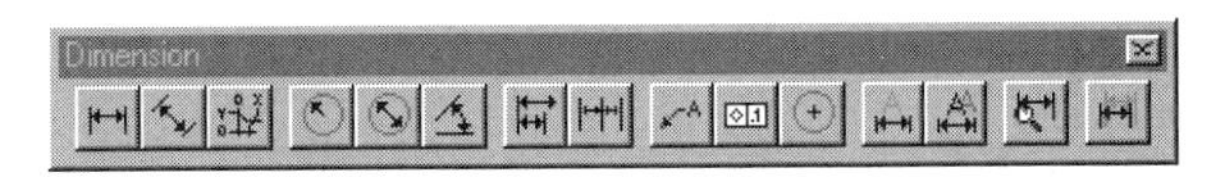

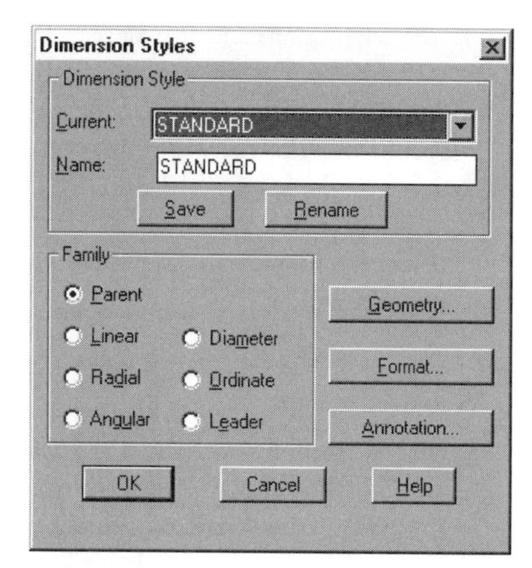

Figure 8.2.

Before moving on, look at the three call boxes at the right. We will make changes only in the Annotation . . . subdialog, but you may wish to look at the others while you are here. If you do, simply open the dialog boxes and cancel them when you are through looking. Geometry . . . calls a subdialog that allows you to adjust aspects of the lines, arrowheads, extension lines, and center marks that make up a dimension. Format . . . allows for changes in the placement of dimension text relative to dimension geometry. Annotation . . . allows changes in the text and measurements provided for AutoCAD dimensions.

⌖ **Click on "Annotation".**

This will call the Annotation dialog box shown in *Figure 8.3.* There are areas here for adjusting Primary Units, which is what we are interested in, Tolerance, Alternate Units, and Text style. We will explore tolerances later. Alternate units are dimension units that may be automatically presented in parentheses along with primary units. Text styles you know about from the last chapter. Standard will be the default text style for dimen-

Figure 8.3.

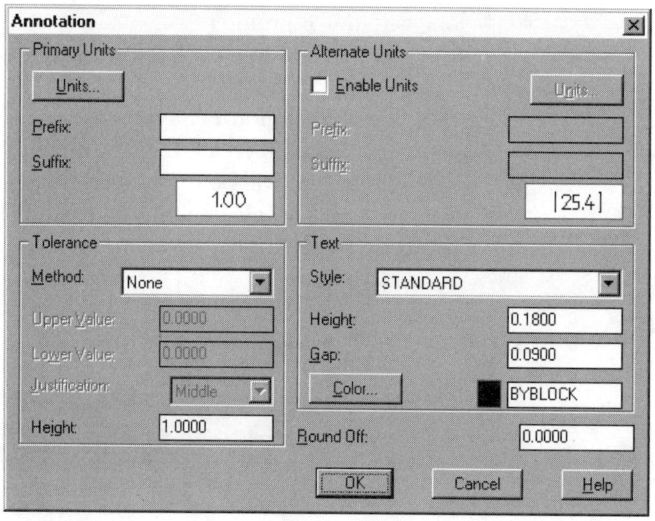

sions, but if you have another text style defined in your drawing, you could select it here. Then all dimensions drawn with this dimension style would have the selected text style.

Click on "Units" in the Primary Units box on the upper left of the dialog box.

This will bring up the Primary Units dialog box shown in *Figure 8.4.* There are adjustments available for units, dimension precision, angular units, and tolerances. The lists under Units and Angles are the same lists used by the UNITS command. The Units should show "Decimal" and the Angles box should show "Decimal Degrees". If for any reason these are not showing in your box you should make these changes now. For our purposes all we need to change is the number of decimal places showing in the Dimension Precision box. By default it will be 0.0000. We will change it to 0.00.

Click on the arrow to the right of the Precision box.

This will open a list of precision settings ranging from 0 to 0.00000000.

Click on 0.00.

This will close the list box and show 0.00 as the selected precision. At this point we are ready to complete this part of the procedure by returning to the Dimension Style dialog box and saving our changes.

Click on "OK" to exit Primary Units.

Click on "OK" to exit Annotation.

Click on "Save" to save the change in dimension precision.

This will save the change in dimension style B. At this point dimension style B has all the AutoCAD default settings except that all dimensions, linear and angular, will be shown as two-place decimals. Next we will set the angular precision separately to 0 places.

Click on the Angular radio button.

Click on Annotation.

Click on Units.

Change Dimension Precision to 0 decimal places.

Be sure to make the change in the same Precision box that you used before. It is the one on the left, under Dimension, not the one on the right under Tolerance. The two boxes are identical and since the Tolerance box is directly under the Angle box it may seem natural to make the change here. But the Tolerance box only changes the

Figure 8.4.

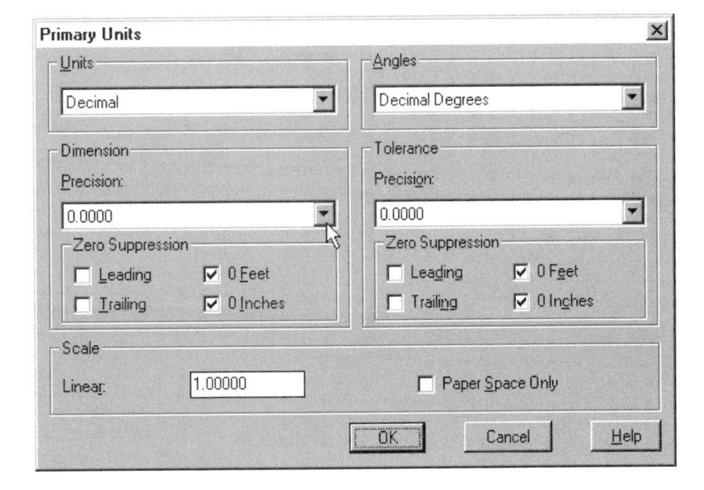

precision of tolerances, which we are not using right now. The Dimension box is used to adjust the precision of all types of dimensions.

⌗ **Click on "OK" to exit Primary Units.**

⌗ **Click on "OK" to exit Annotation.**

⌗ **Click on "Save" to save the changes to B.**

⌗ **Click on "OK" to exit the Dimension Styles dialog box.**

⌗ **Finally, go to the File menu, save template drawing B, and use New to open a new drawing.**

If the new drawing is opened with the B prototype, the B dimension style will be current for the next task.

8.2 DRAWING LINEAR DIMENSIONS

AutoCAD Release 14 has many commands and features that aid in the drawing of dimensions, as evidenced by the fact that there is an entire pull down menu for dimensioning only. In this exercise you will create some basic linear dimensions in the now-current B style.

⌗ **To prepare for this exercise, draw a triangle (ours is 3.00, 4.00, 5.00) and a line (6.00) above the middle of the display, as shown in *Figure 8.5.***

Exact sizes and locations are not critical.. We will begin by adding dimensions to the triangle.

The dimensioning commands are streamlined and efficient. Their full names, however, are all rather long. They all begin with "dim" and are followed by what used to be the name of an option. For example: DIMLINEAR, DIMALIGNED, and DIMANGULAR. They do have somewhat shorter aliases, which we will show you as we go along. We encourage you to use the Dimensioning menu or the Dimensioning toolbar to avoid typing these names.

We begin by placing a linear dimension below the base of the triangle.

⌗ **Type "dimlin", select Linear from the Dimension menu, or the Linear Dimension tool from the Dimension toolbar.**

Figure 8.5.

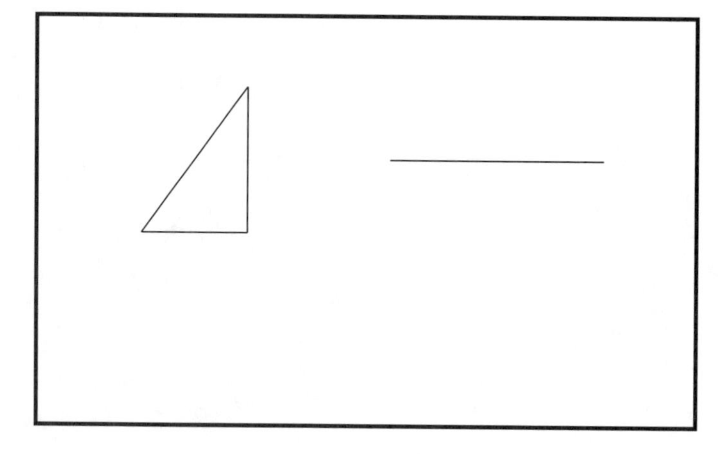

This will initiate the DIMLINEAR command, with the following prompt appearing in the command area:

First extension line origin or press ENTER to select:

There are two ways to proceed at this point. One is to show where the extension lines should begin, and the other is to select the base of the triangle itself and let AutoCAD position the extension lines. In most simple applications the latter method is faster.

⌖ **Press Enter to indicate that you will select an object.**

AutoCAD will replace the cross hairs with a pickbox and prompt for your selection:

Select object to dimension:

⌖ **Select the horizontal line at the bottom of the triangle, as shown by point 1 in *Figure 8.6.***

AutoCAD immediately creates a dimension, including extension lines, dimension line, and text, that you can drag out away from the selected line. AutoCAD will place the dimension line and text where you indicate, but will keep the text centered between the extension lines. The prompt is as follows:

Dimension line location
(MText/Text/Angle/Horizontal/Vertical/Rotated):

In the default sequence, you will simply show the location of the dimension. If you wish to alter the text you can do so using the "Mtext" or "Text" options, or you can change it later with a command called DIMEDIT. "Angle", "Horizontal", and "Vertical" allow you to specify the orientation of the text. Horizontal text is the default for linear text. "Rotated" allows you to rotate the complete dimension, so that the extension lines move out at an angle from the object being dimensioned (text remains horizontal).

NOTE: If the dimension variable "dimsho" is set to 0 (off), you will not be given an image of the dimension to drag into place. The default setting is 1 (on), so this should not be a problem. If, however, it has been changed in your drawing, type "dimsho" and then "1" to turn it on again.

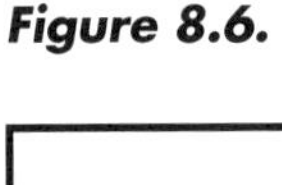

Figure 8.6.

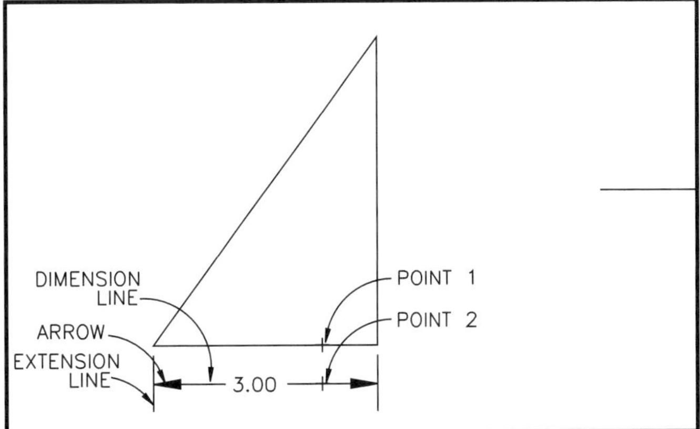

⌗ **Pick a location about .50 below the triangle, as shown by point 2 in *Figure 8.6.***

Bravo! You have completed your first dimension.

Notice that our figure and others in this chapter are shown zoomed in on the relevant object for the sake of clarity. You may zoom or not as you like.

At this point, take a good look at the dimension you have just drawn to see what it consists of. As in *Figure 8.6,* you should see the following components: two extension lines, two "arrows," a dimension line on each side of the text, and the text itself.

Notice also that AutoCAD has automatically placed the extension line origins a short distance away from the triangle base (you may need to zoom in to see this). This distance is controlled by a dimension variable called "dimexo", which can be changed in the Dimension Style dialog box under Geometry. It is one of many variables that control the look of AutoCAD dimensions. Another example of a dimension variable is "dimasz", which controls the size of the arrows at the end of the extension lines.

Next, we will place a vertical dimension on the right side of the triangle. You will see that DIMLINEAR handles both horizontal and vertical dimensions.

⌗ **Repeat the DIMLINEAR command.**

You will be prompted for extension line origins as before:

First extension line origin or press ENTER to select:

This time we will show the extension line origins manually.

⌗ **Pick the right-angle corner at the lower right of the triangle, point 1 in *Figure 8.7.* AutoCAD will prompt for a second point:**

Second extension line origin:

Even though you are manually specifying extension line origins, it is not necessary to show the exact point where you want the line to start. AutoCAD will automatically set the dimension lines slightly away from the line as before, according to the setting of the dimexo dimension variable.

⌗ **Pick the top intersection of the triangle, point 2 in the figure.**

From here on, the procedure will be the same as before. You should have a dimension to drag into place, and the following prompt:

Figure 8.7.

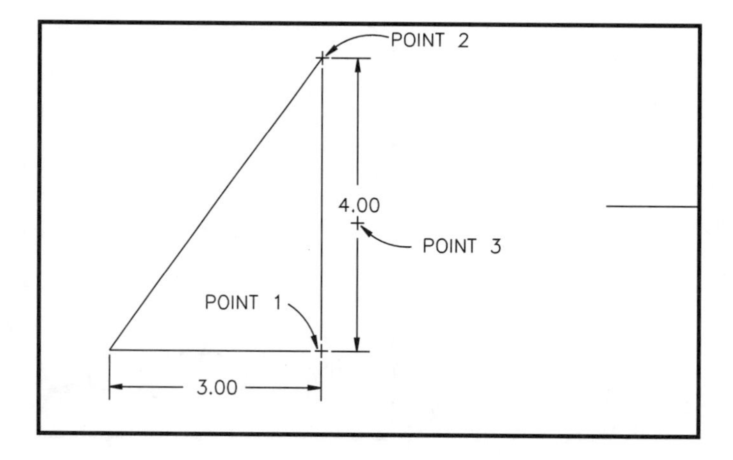

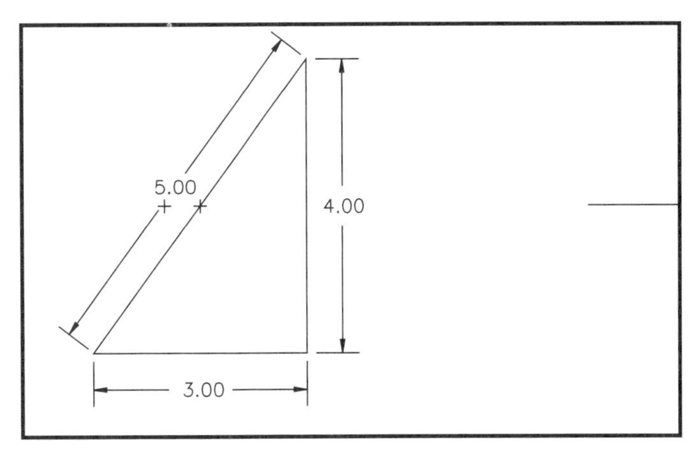

Figure 8.8.

Dimension line location
(Mtext/Text/Angle/Horizontal/Vertical/Rotated):

⌗ Pick a point .50 to the right of the triangle, point 3 in the figure.

Your screen should now include the vertical dimension, as shown in *Figure 8.7.*

Now let's place a dimension on the diagonal side of the triangle. For this we will need the DIMALIGNED command.

⌗ Type "dimali" or select Aligned from the Dimension menu.

⌗ Press Enter, indicating that you will select an object.

AutoCAD will give you the pickbox and prompt you to select an object to dimension.

⌗ Select the hypotenuse of the triangle.

⌗ Pick a point approximately .50 above and to the left of the line.

Your screen should resemble *Figure 8.8.* Notice that AutoCAD retains horizontal text in aligned and vertical dimensions as the default.

8.3 DRAWING MULTIPLE LINEAR DIMENSIONS—BASELINE AND CONTINUE

DIMBASELINE and DIMCONTINUE allow you to draw multiple linear dimensions more efficiently. In baseline format you will have a series of dimensions all measured from the same initial origin. In continued dimensions there will be a string of chained dimensions in which the second extension line for one dimension becomes the first extension for the next.

⌗ To prepare for this exercise, be sure that you have a 6.00 horizontal line as shown in *Figure 8.5* at the beginning of Task 8.2.

Although the figures in this exercise will show only the line, leave the triangle in your drawing, because we will come back to it in the next task. In this exercise we will be placing a set of baseline dimensions on top of the line and a continued series on the bottom. In order to use either Baseline or Continue, you must have one linear dimension already drawn on the line you wish to dimension. So we will begin with this.

⊕ Select Linear from the Dimension menu, or the Linear Dimension tool from the Dimension toolbar.

⊕ Pick the left end point of the line for the origin of the first extension line.

⊕ Pick a second extension origin 2.00 to the right of the first, as shown in *Figure 8.9.*

⊕ Pick a point .50 above the line for dimension text.

You should now have the initial 2.00 dimension shown in *Figure 8.9.* We will use DIMBASELINE to add the other dimensions above it.

⊕ Type "dimbase" or select Baseline from the Dimensioning menu, or the Baseline Dimension tool from the Dimension toolbar.

AutoCAD uses the first extension line origin you picked again and prompts for a second:

Specify a second extension line origin or (Undo/Select):

⊕ Pick a point 4.00 to the right of the first extension line (in other words, 4.00 from the left end of the 6.00 line).

The second baseline dimension shown in Figure 8.9 should be added to your drawing. We will add one more. AutoCAD leaves you in the DIMBASELINE command so that you can add as many baseline dimensions as you want.

⊕ Pick the right end point of the line.

Your screen should resemble *Figure 8.10.* See how quickly AutoCAD draws dimensions? As long as you have your dimension style defined the way you want it, this can make dimensioning go very quickly.

AutoCAD will still be prompting for another second extension line origin. We will need to press Enter twice to exit DIMBASELINE.

⊕ Press Enter to end extension line selection.

The first time you press Enter, AutoCAD will show you a new prompt:

Select base dimension:

This prompt would allow you to begin a new series of baseline dimensions from a dimension other than the last one you drew. Bypassing this prompt will take you back to command prompt.

Figure 8.9.

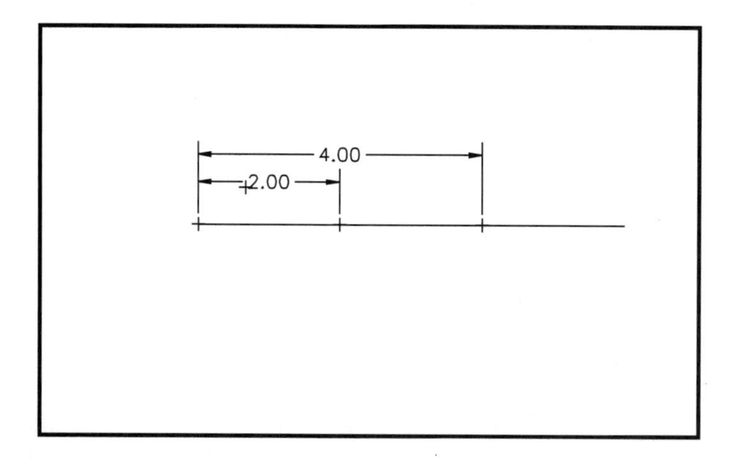

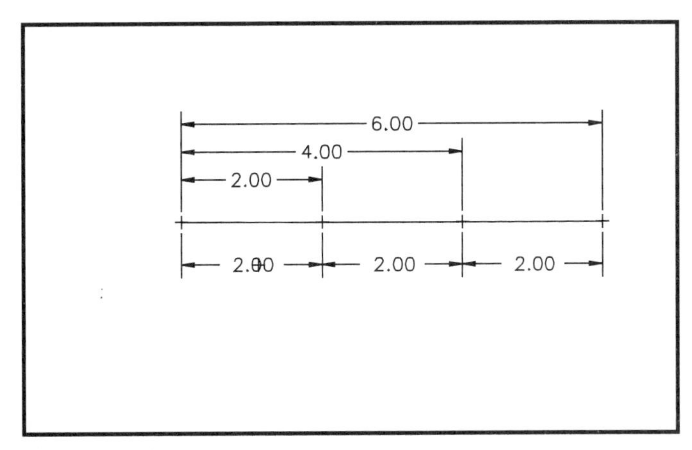

Figure 8.10.

Press Enter to exit DIMBASELINE.

Continued Dimensions

Now we will place three continued dimensions along the bottom of the line, as shown in *Figure 8.11.* The process is very similar to the baseline process so you should need little help at this point.

Begin by using DIMLINEAR to place an initial horizontal dimension .50 below the line, showing a length of 2.00 from the left end, as shown in *Figure 8.11.*

Type "dimcont" or select Continue from the Dimension menu, or the Continue Dimension tool from the Dimension toolbar.

Pick a second extension line origin 2.00 to the right of the last extension line.

Be sure to pick this point on the 6.00 line, otherwise the extension line will be left hanging.

Pick another second extension line at the right end of the line.

Figure 8.11.

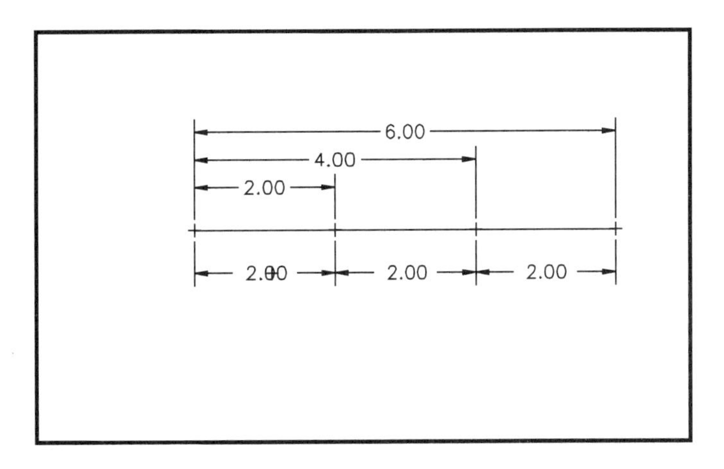

Press Enter to end the continued series.
Press Enter to exit DIMCONTINUE.
Done!

8.4 DRAWING ANGULAR DIMENSIONS

Angular dimensioning works much like linear dimensioning, except that you will be prompted to select objects that form an angle. AutoCAD will compute an angle based on the geometry that you select (two lines, an arc, part of a circle, or a vertex and two points) and construct extension lines, a dimension arc, and text specifying the angle.

For this exercise we will return to the triangle and add angular dimensions to two of the angles as shown in *Figure 8.12.*

Type "dimang" or select Angular from the Dimension menu, or the Angular Dimension tool from the Dimension toolbar.
The first prompt will be:

Select arc, circle, line, or press ENTER:

The prompt shows that you can use angular dimensions to specify angles formed by arcs and portions of circles as well as angles formed by lines. If you press Enter (press ENTER), you can specify an angle manually by picking its vertex and a point on each side of the angle. We will begin by picking lines, the most common use of angular dimensions.

Select the base of the triangle.
You will be prompted for another line:

Second line:

Select the hypotenuse.
As in linear dimensioning, AutoCAD now shows you the dimension lines and lets you drag them into place (assuming dimsho is on). The prompt asks for a dimension arc location and also allows you the option of changing the text or the text angle.

Dimension arc line location (Mtext/Text/Angle):

Figure 8.12.

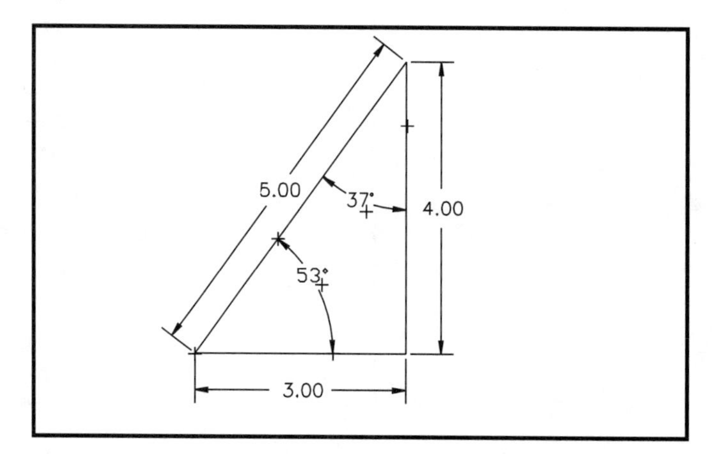

⊕ **Move the cursor around to see how the dimension may be placed and then pick a point between the two selected lines, as shown in *Figure 8.12.***

The lower left angle of your triangle should now be dimensioned, as in *Figure 8.12.* Notice that the degree symbol is added by default in angular dimension text.

We will dimension the upper angle by showing its vertex, using the press ENTER option.

⊕ **Repeat DIMANGULAR.**

⊕ **Press Enter.**

AutoCAD prompts for an angle vertex.

⊕ **Point to the vertex of the angle at the top of the triangle.**

AutoCAD prompts:

First angle end point:

⊕ **Pick a point along the hypotenuse.**

In order to be precise this should be a snap point. The most dependable one will be the lower left corner of the triangle.

AutoCAD prompts:

Second angle end point:

⊕ **Pick any point along the vertical side of the triangle.**

There should be many snap points on the vertical line, so you should have no problem.

⊕ **Move the cursor slowly up and down within the triangle.**

Notice how AutoCAD places the arrows outside the angle when you approach the vertex and things get crowded. Also notice that if you move outside the angle, AutoCAD switches to the outer angle.

⊕ **Pick a location for the dimension arc as shown in *Figure 8.12.***

8.5 DIMENSIONING ARCS AND CIRCLES

The basic process for dimensioning circles and arcs is as simple as those we have already covered. It can get tricky, however, when AutoCAD does not place the dimension where you want it. Text placement can be controlled by adjusting dimension variables. In this exercise we will create a center mark and some diameter and radius dimensions.

⊕ **To prepare for this exercise, draw two circles at the bottom of your screen, as shown in *Figure 8.13.***

The circles we have used have radii of 2.00 and 1.50.

⊕ **Select Center Mark from the Dimension menu, or the Center Mark tool from the Dimension toolbar.**

You cannot use "dimcen" as the command name because that is the name of the dimension variable that controls the style of the center mark. If you are typing you will need to type the whole name "dimcenter".

Center marks resemble blips, but they are actual lines and will appear on a plotted drawing. They are the simplest of all dimension features to create and are created automatically as part of some radius and diameter dimensions.

AutoCAD prompts:

Select arc or circle:

⊕ **Select the smaller circle.**

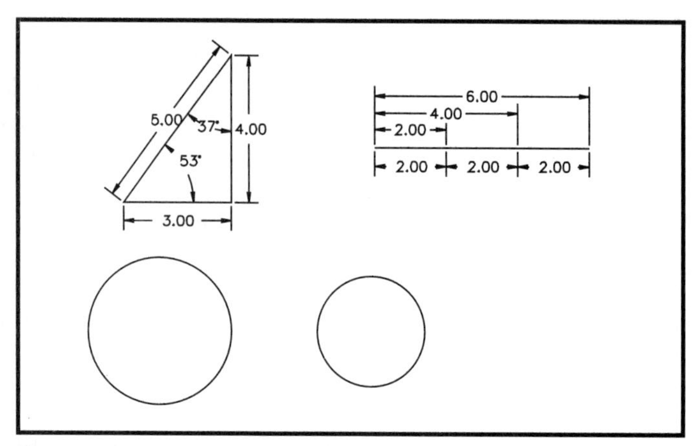

Figure 8.13.

A center mark will be drawn in the 1.50 circle, as shown in *Figure 8.14.*

TIP: A different type of center mark can be produced by changing the dimension variable "dimcen" from .09 to −.09. The result is shown in the dimension variables chart, *Figure 8.16,* at the end of this section.

Now we will add the diameter dimension shown on the larger circle in *Figure 8.14.*

⌖ **Type "dimdia" or select Diameter from the Dimension menu, or the Diameter Dimension tool from the Dimension toolbar.**

AutoCAD will prompt:

Select arc or circle:

⌖ **Select the larger circle.**

AutoCAD will show a diameter dimension and ask for Dimension line location. The Text and Angle options allow you to change the dimension text or put it at an angle. If you move your cursor around you will see that you can position the dimension line anywhere around the circle.

Figure 8.14.

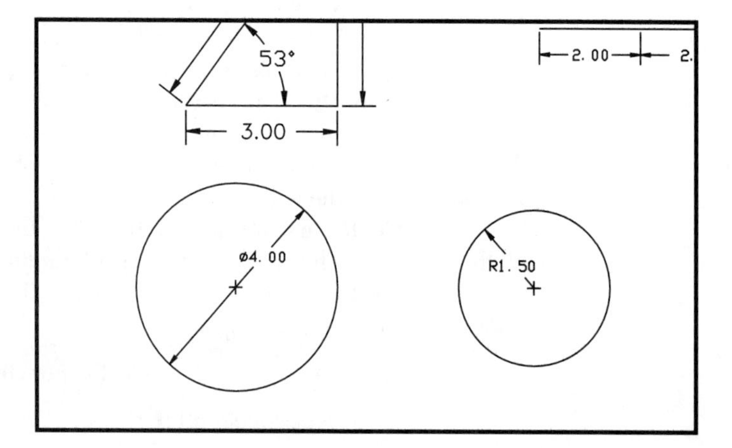

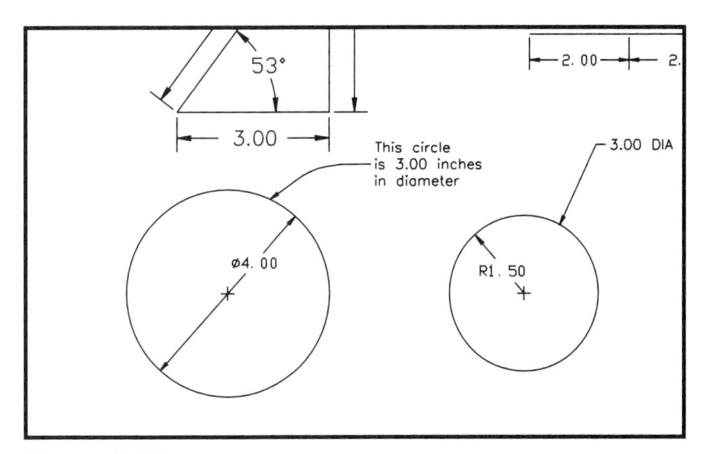

Figure 8.15.

⌖ **Pick a dimension position so that your screen resembles *Figure 8.15.***
Notice that the diameter symbol prefix is added automatically by default.

Radius Dimensions

The procedures for radius dimensioning are exactly the same as those for diameter dimensions. Try adding a radius dimension to the smaller circle, as shown in Figure 8.15.

Dimension Variables

The chart in Figure 8.16 shows some of the more common dimension variables that can be used to change the look and style of dimensions. Dimension variables can be changed through the Dimension Style dialogue box, or by typing the name of the variable at the Command prompt and then entering a new value.

8.6 USING THE BHATCH COMMAND

Automated hatching is another immense timesaver. Release 14 includes two commands for use in cross hatching. Of the two, BHATCH and HATCH, BHATCH is more powerful and easier to use. BHATCH differs from HATCH in that it automatically defines the nearest boundary surrounding a point you have specified.

⌖ **To prepare for this exercise, clear your screen of all previously drawn objects and then draw three rectangles, one inside the other, with the word "TEXT" at the center, as shown in *Figure 8.17.***

⌖ **Type "bh" or select Hatch from the Draw menu.**

There is also a Hatch tool on the Draw toolbar, as shown in *Figure 8.18,* but you may have to move the toolbar into a floating position in order to access it.

All of these methods initiate the BHATCH command, which calls the Boundary Hatch dialog box shown in *Figure 8.19.* At the top of the box you will see the Pattern Type box. Before we can hatch anything we need to specify a pattern. Later we will show AutoCAD what we wish to hatch using the Pick Points method.

The Pattern type currently shown in the image box is a predefined pattern. There are about 68 of these patterns. But we will first use a simple user-defined pattern of straight lines on a 45 degree angle.

⌖ **Click on the arrow to the right of "Predefined".**

COMMONLY USED DIMENSION VARIABLES					
Variable	Default Value	Appearance	DESCRIPTION	New Value	Appearance
dimaso	on	All parts of dim are one entity	Associative dimensioning	off	All parts of dim are separate entities
dimscale	1.00	2.00	Changes size of text & arrows, not value	2.00	2.00
dimasz	.18		Sets arrow size	.38	
dimcen	.09		Center mark size and appearance	-.09	
dimdli	.38		Spacing between continued dimension lines	.50	
dimexe	.18		Extension above dimension line	.25	
dimexo	0.06		Extension line origin offset	.12	
dimtp	0.00	1.50	Sets plus tolerance	.01	$1.50^{+0.01}_{-0.00}$
dimtm	0.00	1.50	Sets minus tolerance	.02	$1.50^{+0.00}_{-0.02}$
dimtol	off	1.50	Generate dimension tolerances (dimtp & dimtm must be set) (dimtol & dimlim cannot both be on)	on	$1.50^{+0.01}_{-0.02}$
dimlim	off	1.50	Generate dimension limits (dimtp & dimtm must be set) (dimtol & dimlim cannot both be on)	on	1.51 1.48
dimtad	off	1.50	Places text above the dimension line	on	1.50
dimtxt	.18	1.50	Sets height of text	.38	1.50
dimtsz	.18	1.50	Sets tick marks & tick height	.25	1.50
dimtih	on	1.50	Sets angle of text When off rotates text to the angle of the dimension	off	1.50
dimtix	off	⌀0.71	Forces the text to inside of circles and arcs. Linear and angular dimensions are placed inside if there is sufficient room	on	⌀0.71

Figure 8.16.

This opens a list including Predefined, User-defined and Custom patterns.

Click on "User-defined".

When you select a user-defined pattern there will be nothing shown in the image box. To create a user-defined pattern you will need to specify an angle and a spacing.

Double click in the Angle edit box, and then type "45".

Double click in the Spacing edit box, and then type ".5".

The remaining box in this area is Double. Double hatching creates double hatch lines running perpendicular to each other according to the specified angle.

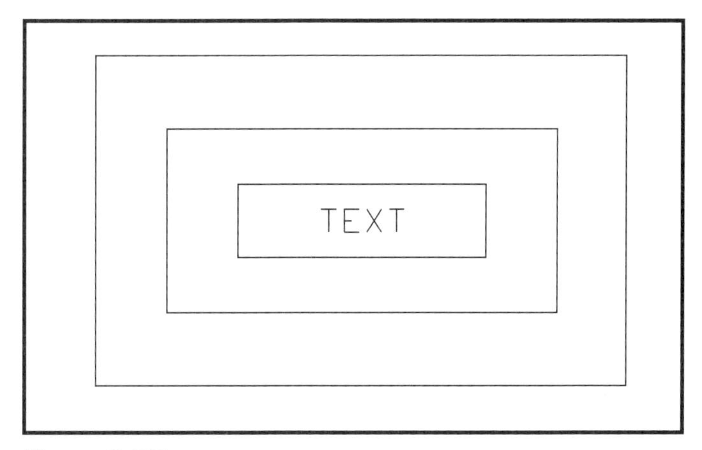

Figure 8.17.

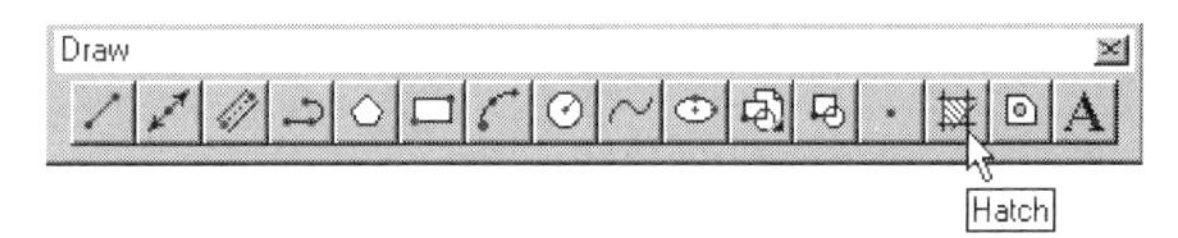

Figure 8.18.

Next we need to show AutoCAD where to place the hatching. To the right in the dialog box is the Boundary area. The first two options are "Pick Points" and "Select Objects". Using the "Pick Points" option, you can have AutoCAD locate a boundary when you point to the area inside it. The "Select Objects" option can be used to create boundaries in the way that the older HATCH command works, by selecting entities that lie along the boundaries.

⌖ **Click on "Pick Points <".**

Figure 8.19.

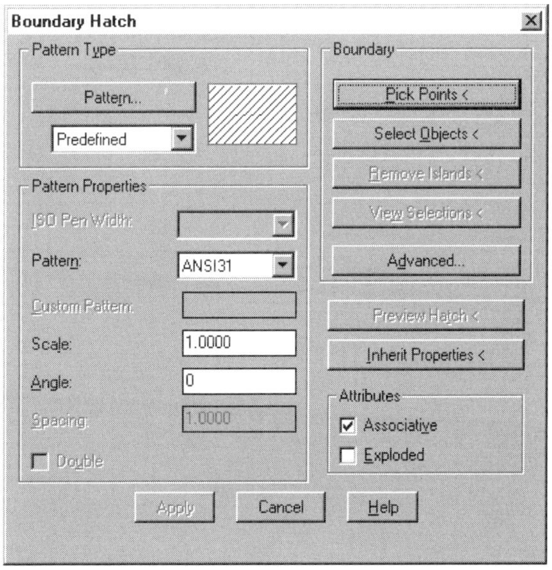

The dialog box will disappear temporarily and you will be prompted as follows:

Select internal point:

Pick a point inside the largest, outer rectangle, but outside the smaller rectangles.

AutoCAD displays these messages, though you may have to press F2 to see them:

Selecting everything visible . . .
Analyzing the selected data . . .
Analyzing internal islands . . .

In a large drawing this process can be time consuming, as the program searches all visible entities to locate the appropriate boundary. When the process is complete, all of the rectangles and the text will be highlighted. It will happen very quickly in this case.

AutoCAD continues to prompt for internal points so that you can define multiple boundaries. Let's return to the dialog box and see what we've done so far.

Press Enter to end internal point selection.

The dialog box will reappear. You may not have noticed, but several of the options that were "grayed out" before are now accessible. We will make use of the "Preview Hatch" option. This allows us to preview what has been specified without leaving the command, so that we can continue to adjust the hatching until we are satisfied.

Click on "Preview Hatch <".

Your screen should resemble *Figure 8.20,* except that you will have to move the Boundary Hatch - Continue box away from the center of the display as we have done. This demonstrates the "normal" style of hatching in which multiple boundaries are hatched or left clear in alternating fashion, beginning with the outermost boundary and working inward. You can see the effect of the other styles by looking at the Advanced Options dialog box. Notice that BHATCH has recognized the text as well as the other interior boundaries.

Click on Continue to return to the Boundary Hatch dialog box.

Figure 8.20.

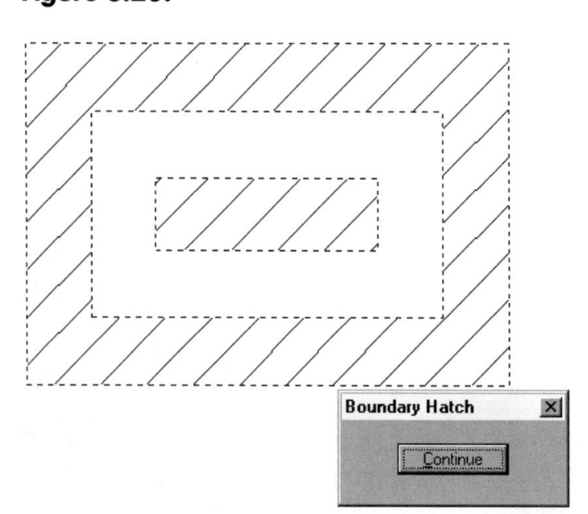

You could complete the hatching operation at this point by clicking on Apply, but let's take a look at a few more details while we are here.

Click on "Advanced".

This calls the Advanced Options dialog box shown in *Figure 8.21.* In the middle you will see the hatching style area. There are three basic options that determine how BHATCH will treat interior boundaries. "Normal" is the current style.

Click on the arrow to the right of "Normal".

You will see the other two styles listed. "Normal" hatches alternate areas moving inward, "Outer" hatches only the outer area, and "Ignore" hatches through all interior boundaries. There is an image box that shows an example of the current style, as shown in *Figure 8.21.*

Click on "Outer" and watch the image box.

It now shows the "Outer" style in which only the area between the outermost boundary and the first inner boundary is hatched.

Click on the arrow again and then on "Ignore" and watch the image box.

In the "Ignore" style, all interior boundaries are ignored and the entire area including the text is hatched.

You can see the same effects in your own drawing, if you like, by clicking on "OK" and then "Preview Hatch" after changing from the "Normal" style to either of the other styles. Of course, the three styles are indistinguishable if you do not have boundaries within boundaries to hatch.

When you are done experimenting, you should select "Normal" style hatching again and return to the Boundary Hatch dialog box.

Now let's take a look at some of the fancier stored hatch patterns that AutoCAD provides.

Click on the arrow to the right of "User-Defined" and select "Predefined".

Now click on the "Pattern".

This opens a long list of AutoCAD's predefined hatch patterns along with sample images on the right. To produce the hatched image in *Figure 8.22,* we chose the Escher pattern.

Move down the list until you come to "Escher".

Select "Escher".

Figure 8.21.

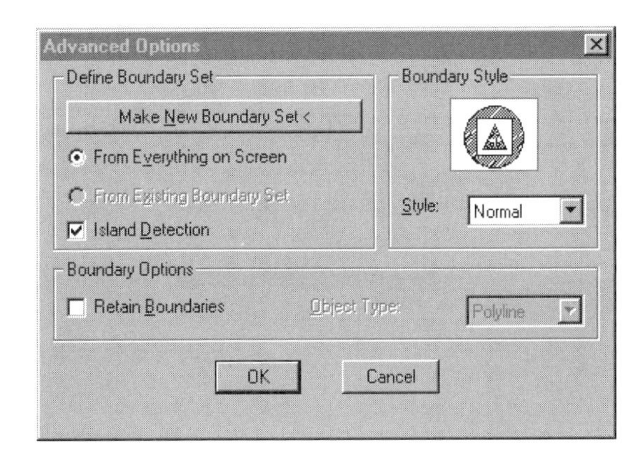

Figure 8.22.

⌗ **Click on Okay to exit the Pattern dialog box.**

To produce the figure we also used a larger scale in this hatching.

⌗ **Double click in the "Scale:" edit box and type "1.5".**

⌗ **Double click in the "Angle:" edit box and type "0".**

⌗ **Click on "Preview Hatch <".**

Your screen should resemble *Figure 8.22*, but remember this is still just the preview. When you are through adjusting hatching, you have the choice of canceling the BHATCH command, so that no hatching is added to your drawing, or completing the process by clicking on "Apply". Apply will confirm the most recent choices of boundaries, patterns, and scales, making them part of your drawing.

⌗ **Click on Continue.**

⌗ **Click on "Apply" to exit BHATCH and confirm the hatching.**

8.7 REVIEW MATERIAL

Questions

1. You are working in a drawing with units set to architectural but when you begin dimensioning, AutoCAD provides 4 place decimal units. What is the problem? What do you need to do so that your dimensioning units match your drawing units?
2. Describe at least one way to change the size of the arrowheads in the dimensions of a drawing.
3. What must you do first before you can create either baseline or continued dimensions?
4. Describe an efficient method for moving dimension text away from the center of a dimensioned circle.
5. What is a "nearest" object snap and why is it important when dimensioning with leaders?
6. What is the difference between HATCH and BHATCH?

COMMANDS

Draw
BHATCH

Dimensioning
DDIM
DIMALIGNED
DIMANGULAR
DIMBASELINE
DIMCONTINUE
DIMLINEAR
DIMSTYLE

Drawing Problems

1. Create a new dimension style called "Dim-2". Dim-2 will use architectural units with 1/2" precision for all units except angles, which will use 2-place decimals. Text in Dim-2 will be .5 units high.
2. Draw an isoceles triangle with vertexes at (4,3), (14,3), and (9,11). Draw a 2" circle centered at the center of the triangle.
3. Dimension the base and one side of the triangle using the Dim-2 dimenson style.
4. Add a diameter dimension to the circle.
5. Add an angle dimension to one of the base angles of the triangle. Make sure that the dimension is placed outside of the triangle.
6. Hatch the area inside the triangle and outside the circle using the predefined "cross" hatch pattern.

PROFESSIONAL SUCCESS

Using AutoCAD in Other Disciplines

Any recent release of AutoCAD is capable of producing the two- and three-dimensional engineering drawings described in this book. The basic release of AutoCAD is geared towards mechanical engineering, and includes symbols and primitives (e.g. dimensioning, line-type, cross-hatch) which were designed to conform with accepted engineering standards. Other types of drawings, including architectural studies, electrical layouts, maps, and piping diagrams, may be produced by using the various extensions and companions to AutoCAD which are available. These extend the capabilities of the basic AutoCAD package, and allow industry-specific primitives and symbols to be specified directly.

One available extension to AutoCAD is AutoCAD Map, an add-on that allows maps and topological diagrams to be easily produced from within AutoCAD release 14. Instead of constructing basic map symbols and geographic features using the geometric primitives (like line or circle) available in the standard AutoCAD package, AutoCAD map allows basic mapping primitives to be added directly to your drawing. This makes creating specialized drawings much easier, and helps to ensure that the symbols and layouts used are based on accepted industry standards.

There is most likely software available that will allow you to use AutoCAD to create the specialized drawings required by your specific field of study. Information about these is available through the AutoDesk web-site (http://www.autodesk.com), which contains complete descriptions on the available packages, as well as demonstrations which may be downloaded and tested.

DRAWING 8—1

Flanged Wheel

Most of the objects in this drawing are straightforward. The keyway is easily done using the TRIM command. Use DIMTEDIT or grips to move the diameter dimension as shown in the reference following.

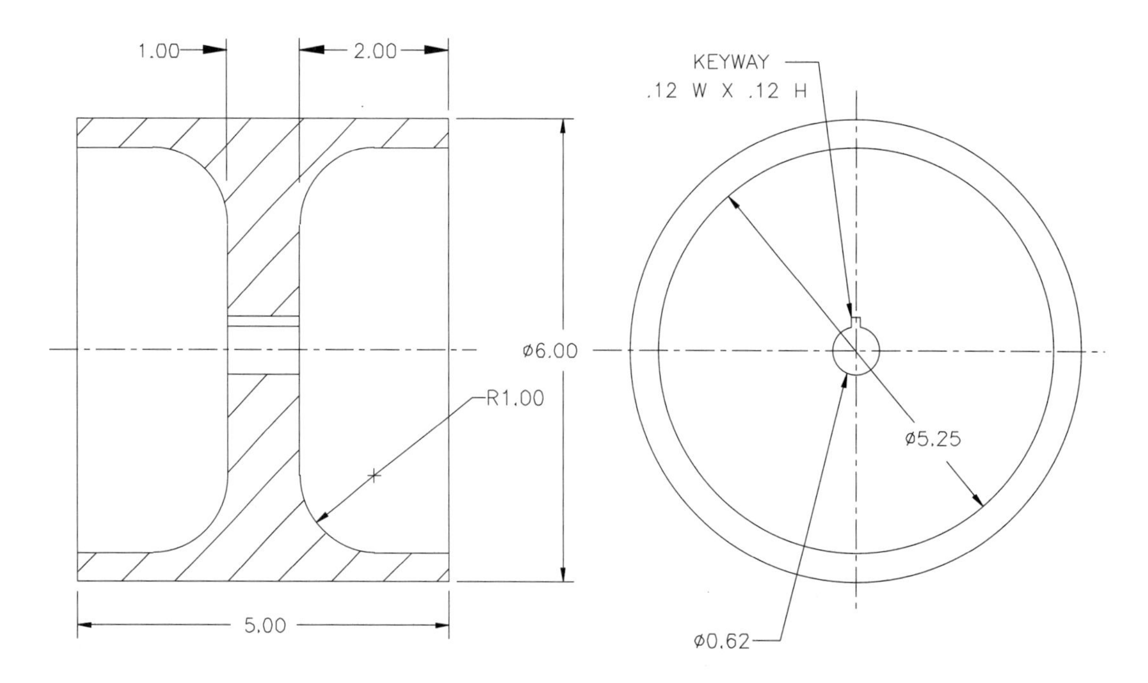

FLANGED WHEEL

Drawing 8–1

Drawing Suggestions

GRID = .25
SNAP = .0625
HATCH line spacing = .50

- You will need a .0625 snap to draw the keyway. Draw a .125 × .125 square at the top of the .63 diameter circle. Drop the vertical lines down into the circle so they may be used to TRIM the circle. TRIM the circle and the vertical lines, using a window to select both as cutting edges.
- Remember to set to layer "hatch" before hatching, layer "text" before adding text, and layer "dim" before dimensioning.

DRAWING 8—2

Shower Head

This drawing makes use of the procedures for hatching and dimensioning you learned in the last two drawings. In addition, it uses an angular dimension, baseline dimensions, leaders, and "%%c" for the diameter symbol.

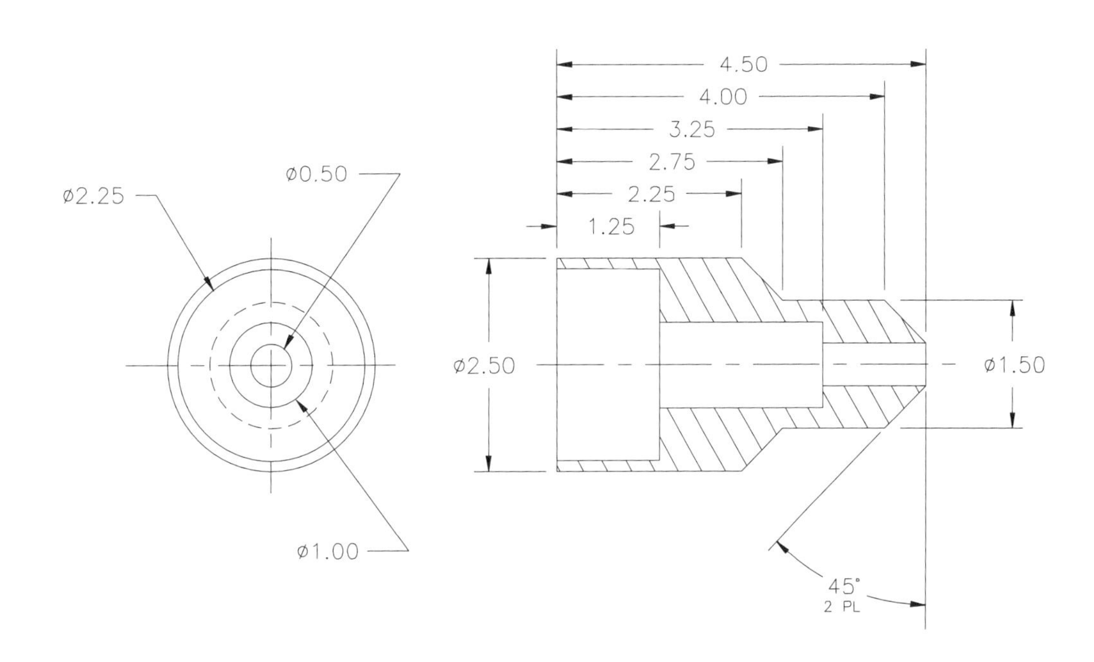

SHOWER HEAD

Drawing 8-2

Drawing Suggestions

GRID = .50
SNAP = .125
HATCH line spacing = .25

- You can save some time on this drawing by using MIRROR to create half of the right side view. Notice, however, that you cannot hatch before mirroring, because the mirror command will reverse the angle of the hatch lines.
- To achieve the angular dimension at the bottom of the right side view, you will need to draw the vertical line coming down on the right. Select this line and the angular line at the right end of the shower head, and the angular extension will be drawn automatically. Add the text "2 PL" using the DIMEDIT command.
- Notice that the diameter symbols in the vertical dimensions at each end of the right side view are not automatic. Use %%c to add the diameter symbol to the text.

DRAWING 8—3

Plot Plan

This architectural drawing makes use of three hatch patterns and several dimension variable changes. Be sure to make these settings as shown. Notice that we have simplified the format of the drawing page for this drawing. This is because the drawings are becoming more involved and because you should need less information to complete them at this point. We will continue to show drawings this way for the remainder of the book.

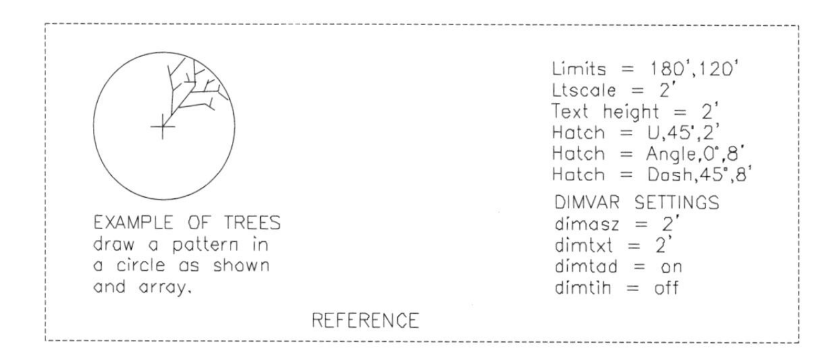

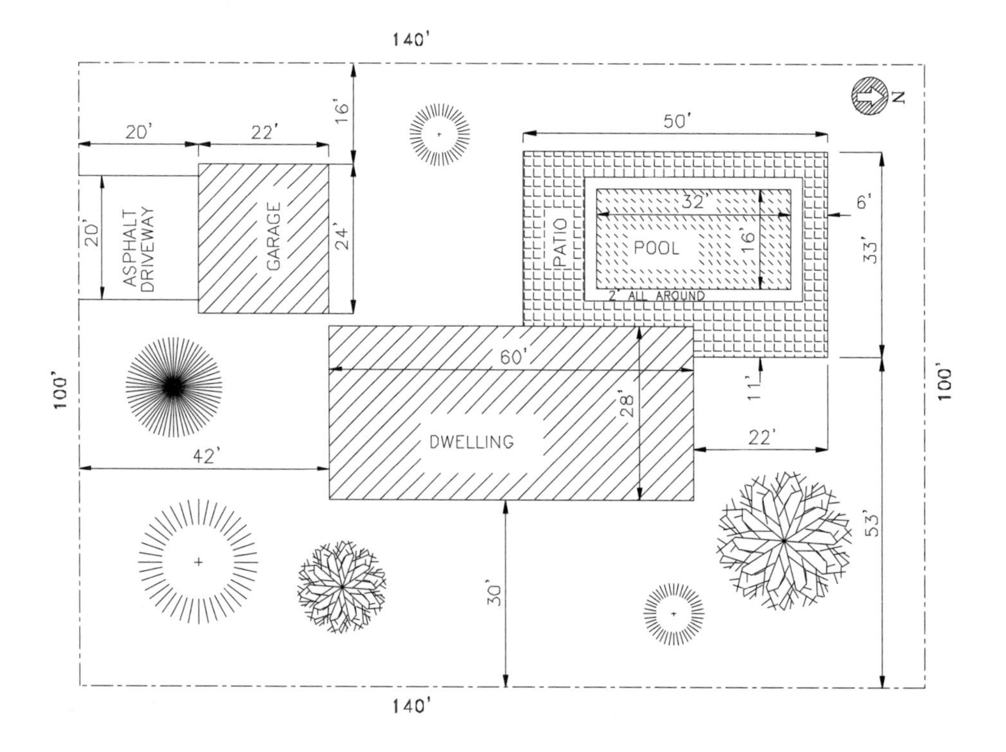

PLOT PLAN

Drawing 8–3

Drawing Suggestions

GRID = 10' LIMITS = 180',120'
SNAP = 1' LTSCALE = 2'

- The "trees" shown here are symbols for oaks, willows, and evergreens.
- BHATCH will open a space around text inside a defined boundary; however, sometimes you will want more white space than BHATCH leaves. The simple solution is to draw a box around the text area as an inner boundary. If the BHATCH style is set to "Normal" it will stop hatching at the inner boundary. Later you can erase the box, leaving an island of white space around the text.

9

Wire Frame and Surface Models

OVERVIEW

It is now time to begin thinking in three dimensions. 3D drawing in AutoCAD is logical and efficient. You can create wireframe models, surface models, or solid models and display them from multiple points of view. In this chapter we will focus on User Coordinate Systems, 3D viewpoints, and the three types of 3D models. These are the primary tools you will need to understand how AutoCAD allows you to work in three dimensions on a two-dimensional screen.

SECTIONS

OBJECTIVES

After reading this chapter, you should be able to:

- Create and view a 3D box and define and save three User Coordinate Systems.
- Use standard edit commands in a UCS and use multiple tiled viewports.
- Create flat surfaces with 3DFACE and remove hidden lines with HIDE.
- Use 3D polygon meshes.
- Create solid boxes and wedges and create the union of two solids.
- Work above the XY plane with ELEV.
- Create the subtraction of two solids and SHADE and RENDER solid objects.
- Draw a clamp, REVSURF designs, and a bushing mount.

9.1 CREATING AND VIEWING A 3D WIREFRAME BOX

In this task we will create a simple 3D box that we can edit in later tasks to form a more complex object. In Chapter 1 we showed how to turn the coordinate system icon (see *Figure 9.1*) off and on. For drawing in 3D you will definitely want it on. If your icon is not visible follow this procedure to turn it on:

1. Open the View menu, highlight "Display", and then "UCS", as shown in *Figure 9.2*
2. Select "on" from the sub menu.

For now, simply observe the icon as you go through the process of creating the box, and be aware that you are currently working in the same coordinate system that you have always used in AutoCAD. It is called the World Coordinate System, to distinguish it from others you will create yourself beginning in Section 9.2.

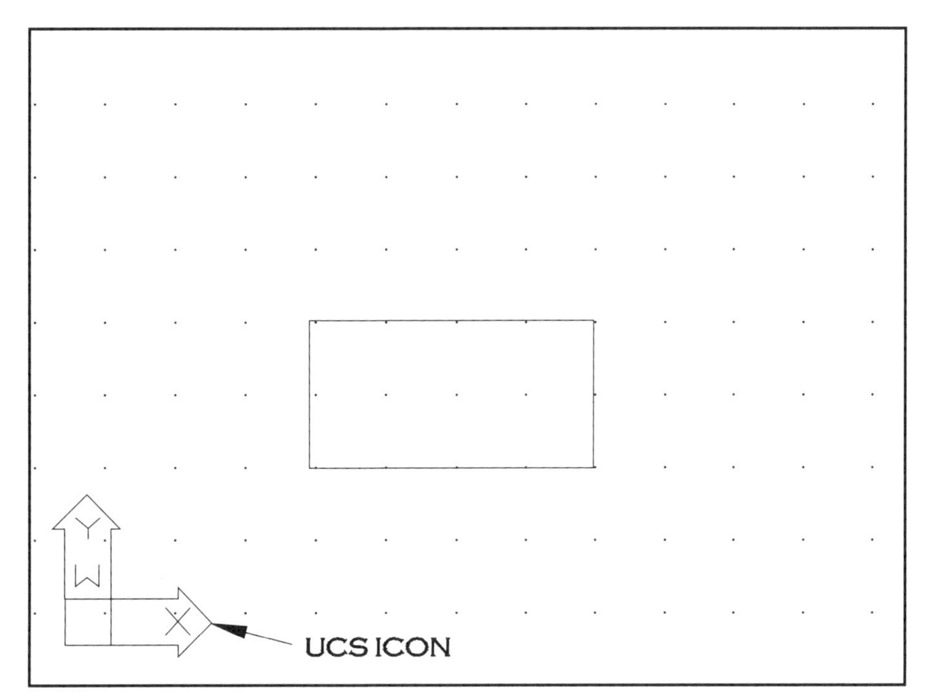

Figure 9.1.

Figure 9.2.

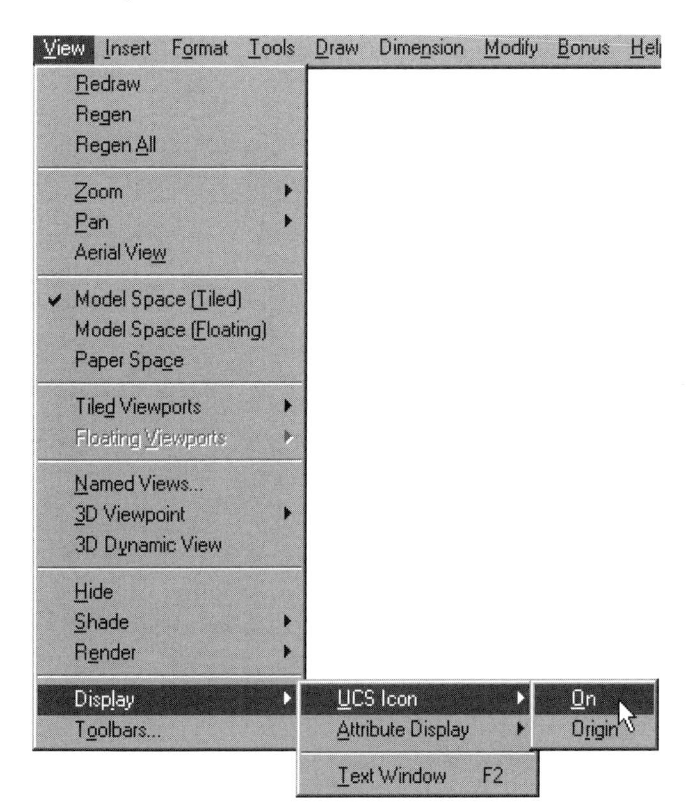

Currently, the origin of the WCS is at the lower left of your screen. This is the point (0,0,0) when you are thinking 3D, or simply (0,0) when you are in 2D. x coordinates increase to the right horizontally across the screen, and y coordinates increase vertically up the screen as usual. The z axis, which we have ignored up until now, currently extends out of the screen towards you and perpendicular to the x and y axes. This orientation of the three planes is also called a plan view. Soon we will switch to a "front, right, top" or "southeast isometric" view.

Let's begin.

⌗ Draw a 2.00 by 4.00 rectangle near the middle of your screen, as shown in *Figure 9.1.* Do not use the RECTANG command to draw this figure because we will want to select individual line segments later on.

Changing Viewpoints

In order to move immediately into a 3D mode of drawing and thinking, our first step will be to change our viewpoint on this object. In Release 14 the simplest and most efficient method is to use the 3D Viewpoints Presets submenu from the View pull down menu. For our purposes, this is the only method you will need.

⌗ Open the View menu and highlight "3D Viewpoint".

This will open a submenu beginning with "Plan View", as shown in *Figure 9.3.* Using this method, AutoCAD will take you directly into 1 of 11 preset viewpoints. In all of the nonisometric views in the submenu (Plan View, Top, Bottom, Left,

Figure 9.3.

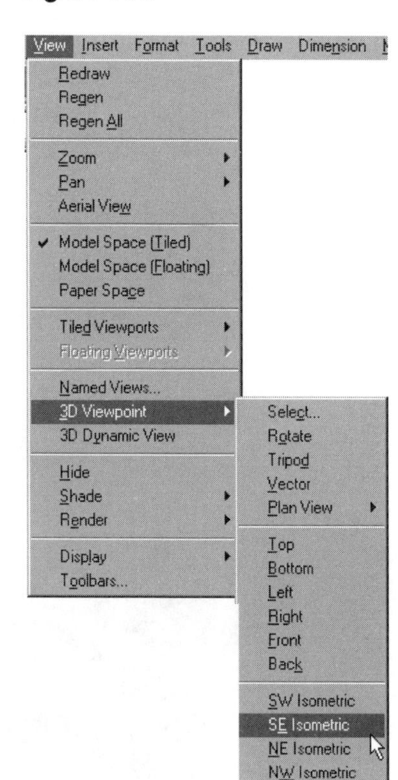

Right, Front, Back) the x and y axes will be rotated 90 degrees or not at all. You will see the drawing from directly above (Plan and Top) or directly below (Bottom), or you will look in along the x axis (Left and Right) or along the y axis (Front and Back).

We will use a Southeast Isometric view. The isometric views all rotate the xy axes plus or minus 45 degrees and take you up 30 degrees out of the XY plane. It is simple if you imagine a compass. The lower right quadrant is the southeast. In a southeast isometric view you will be looking in from this quadrant and down at a 30 degree angle. Try it.

☩ **Select "SE Isometric" from the submenu.**

The menu will disappear and the screen will be redrawn to show the view shown in *Figure 9.4*. Notice how the grid and the coordinate system icon have altered to show our current orientation. These visual aids are extremely helpful in viewing 3D objects on the flat screen and imagining them as if they were positioned in space.

At this point you may wish to experiment with the other views in the 3D Viewpoint submenu. You will probably find the isometric views most interesting at this point. Pay attention to the grid and the icon as you switch views. Variations of the icon you may encounter here and later on are shown in *Figure 9.5*. With some views you will have to think carefully and watch the icon to understand which way the object is being presented.

When you are done experimenting, be sure to return to the southeast isometric view shown in *Figure 9.4*. We will use this view frequently throughout this chapter and the next.

Whenever you change viewpoints, AutoCAD displays the drawing extents, so that the object fills the screen and is as large as possible. Often you will need to zoom out a bit to get some space to work in. This is easily done using the "Scale(X)" option of the ZOOM command.

☩ **Open the View menu, highlight Zoom, and select Scale, or select the Zoom Scale tool from the Zoom flyout on the Standard toolbar, as shown in *Figure 9.6*.**

☩ **Type ".5".**

Figure 9.4.

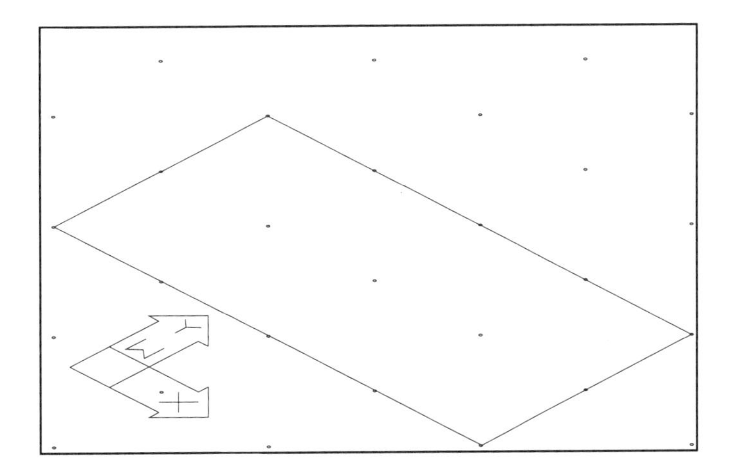

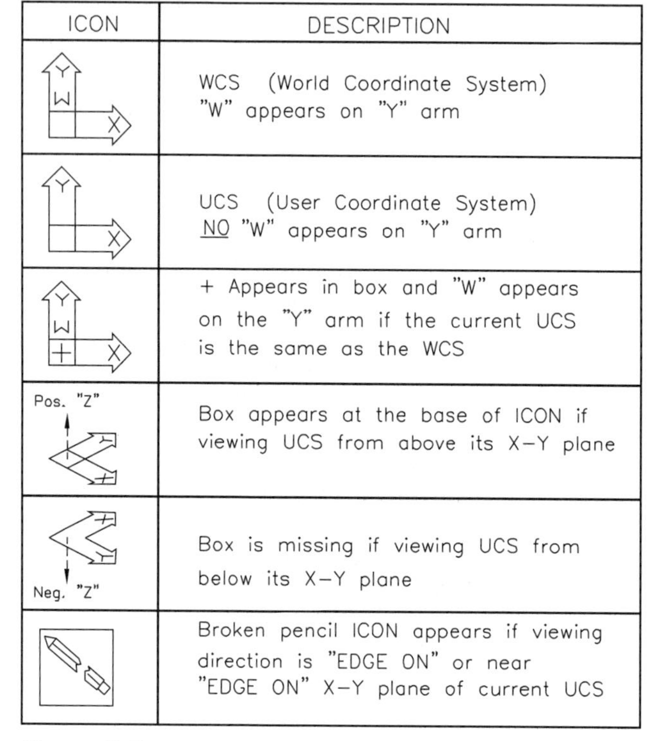

ICON	DESCRIPTION
	WCS (World Coordinate System) "W" appears on "Y" arm
	UCS (User Coordinate System) NO "W" appears on "Y" arm
	+ Appears in box and "W" appears on the "Y" arm if the current UCS is the same as the WCS
Pos. "Z"	Box appears at the base of ICON if viewing UCS from above its X-Y plane
Neg. "Z"	Box is missing if viewing UCS from below its X-Y plane
	Broken pencil ICON appears if viewing direction is "EDGE ON" or near "EDGE ON" X-Y plane of current UCS

Figure 9.5.

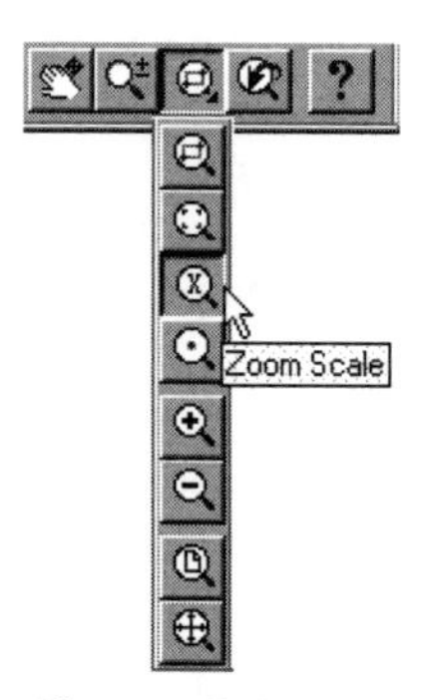

Figure 9.6.

This tells AutoCAD to adjust and redraw the display so that objects appear half as large as before.

Your screen will be redrawn with the rectangle at half its previous magnification.

Entering 3D Coordinates

Next, we will create a copy of the rectangle placed 1.25 above it. This brings up a basic 3D problem: AutoCAD interprets all pointer device point selections as being in the XY plane, so how does one indicate a point or a displacement in the Z direction? There are three possibilities: typed 3D coordinates, X/Y/Z point filters, and object snaps. Object snap requires an object already drawn above or below the XY plane, so it will be of no use right now. We will use typed coordinates first, then discuss how point filters could be used as an alternative. Later we will be using object snap as well.

3D coordinates can be entered from the keyboard in the same manner as 2D coordinates. Often this is an impractical way to enter individual points in a drawing. However, within COPY or MOVE it provides a simple method for specifying a displacement in the Z direction.

☩ Type "co", select Copy from the Modify menu, or the Copy tool from the Modify toolbar.

AutoCAD will prompt for object selection.

☩ Select the complete rectangle.

☩ Press Enter, the space bar, or the enter equivalent button on your pointing device to end object selection.

AutoCAD now prompts for the base point of a vector or a displacement value:

<Base point or displacement>/Multiple:

Typically, you would respond to this prompt and the next by showing the two end points of a vector. However, we cannot show a displacement in the Z direction by pointing. This is important for understanding AutoCAD coordinate systems. Unless an object snap is used, all points picked on the screen with the pointing device will be interpreted as being in the XY plane of the current UCS. Without an entity outside the XY plane to use in an object snap, there is no way to point to a displacement in the Z direction.

⌗ **Type "0,0,1.25".**

AutoCAD now prompts:

Second point of displacement:

You can type the coordinates of another point, or press Enter to tell AutoCAD to use the first entry as a displacement from (0,0,0). In this case, pressing Enter will indicate a displacement of 1.25 in the Z direction, and no change in X or Y.

⌗ **Press Enter.**

AutoCAD will create a copy of the rectangle 1.25 directly above the original. Your screen should resemble *Figure 9.7.*

Using Object Snap

We now have two rectangles floating in space. Our next job is to connect the corners to form a wireframe box. This is done easily using "Endpoint" object snaps. This is a good example of how object snaps allow us to construct entities not in the XY plane of the current coordinate system.

⌗ **Hold down the shift key and press the enter button on your mouse.**

⌗ **Select "Osnap Settings . . ." from the cursor menu.**

⌗ **Click the check box next to "Endpoint."**

⌗ **Click on OK.**

The running Endpoint object snap is now on and will affect all point selection. You will find that object snaps are very useful in 3D drawing and that Endpoint mode can be used frequently.

Now we will draw some lines.

⌗ **Enter the LINE command and connect the upper and lower corners of the two rectangles, as shown in *Figure 9.8.*** (We have removed the grid for clarity, but you will probably want to leave yours on.)

Figure 9.7.

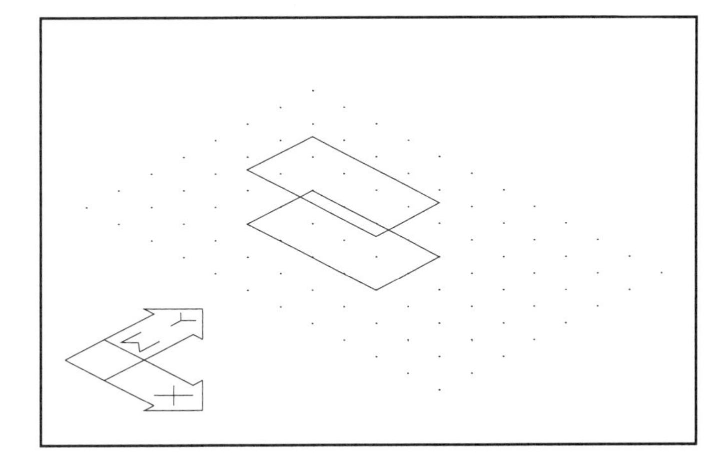

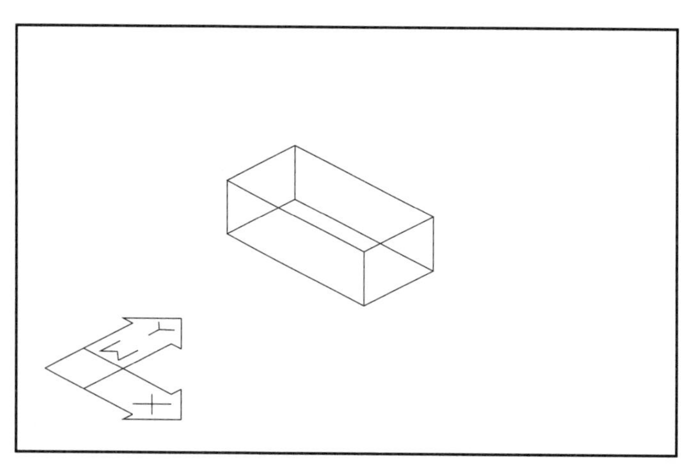

Figure 9.8.

Before going on, pause a moment to be aware of what you have drawn. The box on your screen is a true wireframe model. Unlike an isometric drawing, it is a 3D model that can be turned, viewed, and plotted from any point in space. It is not, however, a solid model or a surface model. It is only a set of lines in 3D space. Removing hidden lines or shading would have no effect on this model.

In the next section, you will begin to define your own coordinate systems that will allow you to perform drawing and editing functions in any plane you choose.

9.2 DEFINING AND SAVING USER COORDINATE SYSTEMS

In this task you will begin to develop new vocabulary and techniques for working with objects in 3D space. The primary tool will be the UCS or DDUCS command. You will also learn to use the UCSICON command to control the placement of the coordinate system icon.

Until now we have had only one coordinate system to work with. All coordinates and displacements have been defined relative to a single point of origin. In Task 1 we changed our point of view, but the UCS icon changed along with it, so that the orientations of the x, y, and z axes relative to the object were retained. With the UCS command you can define new coordinate systems at any point and any angle in space. When you do, you can use the coordinate system icon and the grid to help you visualize the planes you are working in, and all commands and drawing aids will function relative to the new system.

The coordinate system we are currently using is unique. It is called the World Coordinate System and is the one we always begin with. The "W" at the base of the coordinate system icon indicates that we are working in the world system. A User Coordinate System is nothing more than a new point of origin and a new orientation for the x, y, and z axes.

We will begin by defining a User Coordinate System in the plane of the top of the box, as shown in *Figure 9.9.*

⌗ **Leave the Endpoint osnap mode on for this exercise.**
⌗ **Open the Tools menu and highlight "UCS".**

There is also a UCS tool on the UCS toolbar shown in *Figure 9.10,* but you will have to open the toolbar before you can access it. If you type "ucs" or select the tool

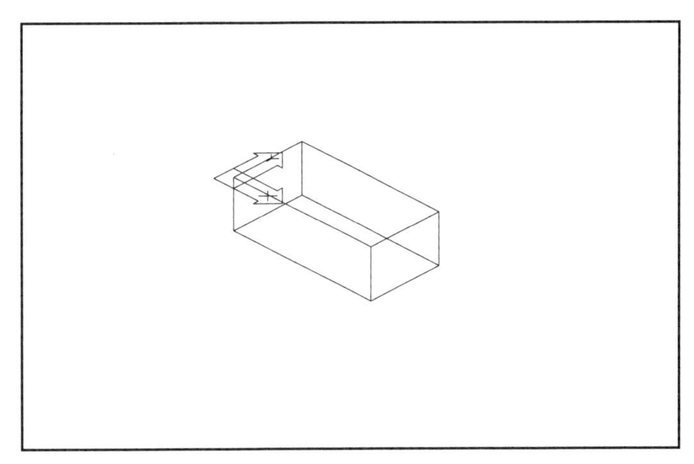

Figure 9.9.

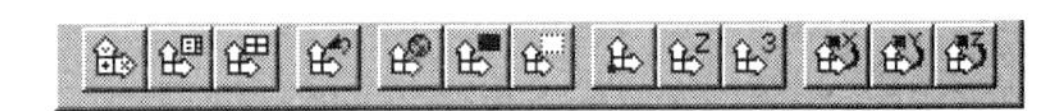

Figure 9.10.

from the toolbar you will work with prompts at the command line. Here we will begin using the pull down menu options and later use the command line as well.

On the pull down menu, you will see options on a submenu, as illustrated in *Figure 9.11*. We will begin by using "Origin" to create a UCS that is parallel to the WCS.

⌗ Select "Origin" from the submenu.

AutoCAD will prompt for a new origin:

Origin point <0,0,0>:

This option does not change the orientation of the three axes. It simply shifts their intersection to a different point in space. We will use this simple procedure to define a UCS in the plane of the top of the box.

⌗ Use the Endpoint object snap to select the top left front corner of the box, as shown by the location of the icon in *Figure 9.9.*

You will notice that the "W" is gone from the icon. However, the icon has not moved. It is still at the lower left of the screen. It is visually helpful to place it at the origin of the new UCS, as in the figure. In order to do this we need the UCSICON command.

⌗ Open the View menu and highlight "Display" and then "UCS Icon".

⌗ Select "Origin".

The icon will move to the origin of the new current UCS, as in *Figure 9.9.* With UCSICON set to origin, the icon will shift to the new origin whenever we define a new UCS. The only exception would be if the origin were not on the screen or too close to an edge for the icon to fit. In these cases the icon would be displayed in the lower left corner again.

The "top" UCS we have just defined will make it easy to draw and edit entities that are in the plane of the top of the box and also to perform editing in planes that are parallel to it, such as the bottom. In the next task we will begin drawing and editing using different coordinate systems, and you will see how this works. For now, we will spend a little more time on the UCS command itself. We will define two more user coordinate systems, but first, let's save this one so that we can recall it quickly when we need it later on.

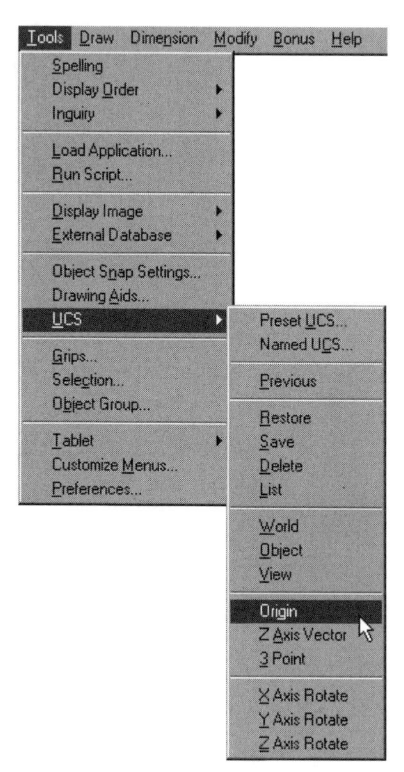

Figure 9.11.

☩ **Open the Tools menu and highlight "UCS".**

This time we will use the "Save" option.

☩ **Select "Save".**

AutoCAD will ask you to name the current UCS so that it can be called out later:

?/Desired UCS name:

We will name our coordinate system "top." It will be the UCS we use to draw and edit in the top plane. This UCS would also make it easy for us to create an orthographic top view later on.

☩ **Type "top".**

The top UCS is now saved and can be recalled using the "Restore" option or by making it "current" using the UCS Control dialog box (select "Named UCS . . ." from the UCS submenu).

Strictly speaking, it is not necessary to save every UCS. However, it often saves time, since it is unlikely that you will have all your work done in any given plane or UCS the first time around. More likely, you will want to move back and forth between major planes of the object as you draw. Be aware also that you can return to the last defined UCS with the "Previous" option.

Next we will define a "front" UCS using the "3point" option.

☩ **Open the Tools menu, highlight "UCS", and select "3 point" from the submenu.**

AutoCAD prompts:

Origin point <0,0,0>:

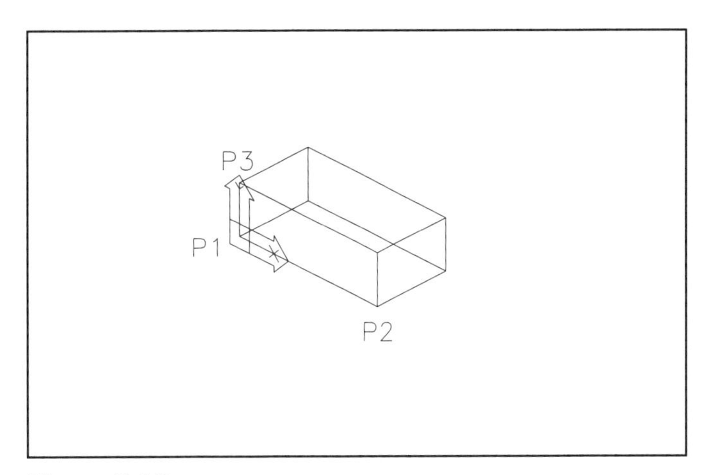

Figure 9.12.

In this option you will show AutoCAD a new origin point, as before, and then a new orientation for the axes as well. Notice that the default origin is the current one. If we retained this origin, we could define a UCS with the same origin and a different axis orientation.

Instead, we will define a new origin at the lower left corner of the front of the box, as shown in *Figure 9.12.*

⌖ **With the Endpoint osnap on, pick P1, as shown in the figure.**

AutoCAD now prompts you to indicate the orientation of the x axis:

Point on positive portion of the X axis <1.00,0.00,–1.25>:

⌖ **Pick the right front corner of the box, P2, as shown.**

The object snap ensures that the new x axis will now align with the front of the object. AutoCAD prompts for the y axis orientation:

Point on positive-Y portion of the UCS XY plane <0.00,1.00,–1.25>:

By definition, the y axis will be perpendicular to the x axis; therefore AutoCAD needs only a point that shows the plane of the y axis and its positive direction. Because of this, any point on the positive side of the y plane will specify the y axis correctly. We have chosen a point that is on the y axis itself.

⌖ **Pick P3, as shown.**

When this sequence is complete, you will notice that the coordinate system icon has rotated along with the grid and moved to the new origin as well. This UCS will be convenient for drawing and editing in the front plane of the box, or editing in any plane parallel to the front, such as the back.

9.3 USING DRAW AND EDIT COMMANDS IN A UCS

Now the fun begins. Using our three new coordinate systems and one more we will define later, we will give the box a more interesting "slotted wedge" shape. In this task we will cut away a slanted surface on the right side of the box. Since the planes we will be working in are parallel to the front of the box, we will begin by making the "front" UCS current. All our work in this task will be done in this UCS.

Look at *Figure 9.13.* We will draw a line down the middle of the front (Line 1) and use it to trim another line coming in at an angle from the right (Line 2).

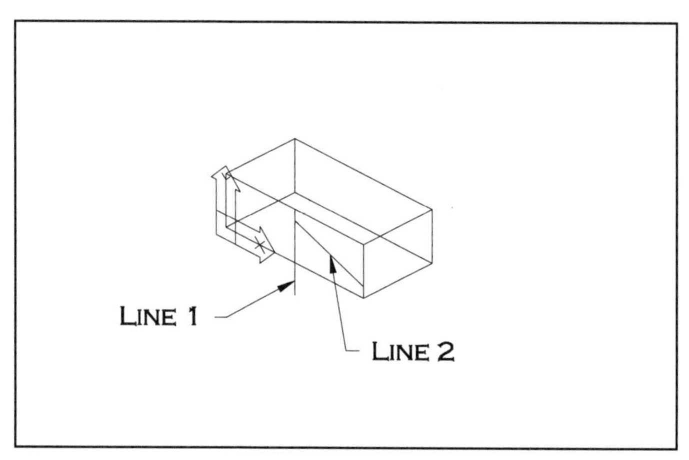

Figure 9.13.

⌗ **Type "L" or select the Line tool from the Draw toolbar.**
⌗ **Open the object snap cursor menu.**
⌗ **Select Midpoint.**
This will temporarily override the Endpoint object snap.
⌗ **Point to the top front edge of the box.**
AutoCAD will snap to the midpoint of the line.
⌗ **Make sure that ortho is on (F8).**
Notice how ortho works as usual, but relative to the current UCS.
⌗ **Pick a second point anywhere below the box.**
⌗ **Exit the LINE command.**

This line will be trimmed later, so the exact length does not matter.

Next we will draw Line 2 on an angle across the front. This line will become one edge of a slanted surface. Your snap setting will need to be at .25 or smaller, and ortho will need to be off. The grid, snap, and coordinate display all work relative to the current UCS, so it is a simple matter to draw in this plane.

⌗ **Check your snap setting and change it to .25 if necessary.**
⌗ **Turn ortho off (F8).**
⌗ **Enter the LINE command.**
⌗ **Hold down the shift key and press the enter button on your cursor to open the object snap cursor menu.**
⌗ **Select None.**
⌗ **Use incremental snap (F9) to pick a point .25 down from the top edge of the box on line 1, as shown in *Figure 9.13.***
⌗ **Hold down the shift key and press the enter button on your cursor to open the object snap cursor menu.**
⌗ **Select None again.**
⌗ **Pick a second point .25 up along the right front edge of the box, as shown.**
⌗ **Exit the LINE command.**
Now trim line 1.
⌗ **Type "tr" or select the Trim tool from the Modify toolbar.**
You will see the following message in the command area:

View is not plan to UCS. Command results may not be obvious.

In the language of AutoCAD 3D, a view is plan to the current UCS if the XY plane is in the plane of the monitor display and its axes are parallel to the sides of the screen. This is the usual 2D view in which the y axis aligns with the left side of the display and the x axis aligns with the bottom of the display. In previous chapters wc always worked in plan view. In this chapter we have not been in plan view since the beginning of Section 9.1

With this message, AutoCAD is warning us that boundaries, edges, and intersections may not be obvious as we look at a 3D view of an object. For example, lines that appear to cross may be in different planes.

Having read the warning, we continue.

⌖ **Select Line 2 as a cutting edge.**

⌖ **Press Enter to end cutting edge selection.**

⌖ **Point to the lower end of Line 1.**

⌖ **Press Enter to exit TRIM.**

Your screen should resemble *Figure 9.14.*

Now we will copy our two lines to the back of the box. Since we will be moving out of the front plane, which is also the XY plane in the current UCS, we will require the use of the running Endpoint object snaps to specify the displacement vector.

⌖ **Type "co" or select the Copy tool from the Modify toolbar.**

⌖ **Pick lines 1 and 2.** (You may need to zoom in on the box at this point to pick Line 1.)

⌖ **Press Enter to end object selection.**

⌖ **Use the Endpoint osnap to pick the lower front corner of the box, P1 as shown in *Figure 9.15.***

⌖ **At the prompt for a second point of displacement, use the ENDpoint osnap to pick the lower right back corner of the box, P2 as shown.**

Your screen should now resemble *Figure 9.15.*

What remains is to connect the front and back of the surfaces we have just outlined and then trim away the top of the box. We will continue to work in the front UCS and to use Endpoint osnaps.

Figure 9.14.

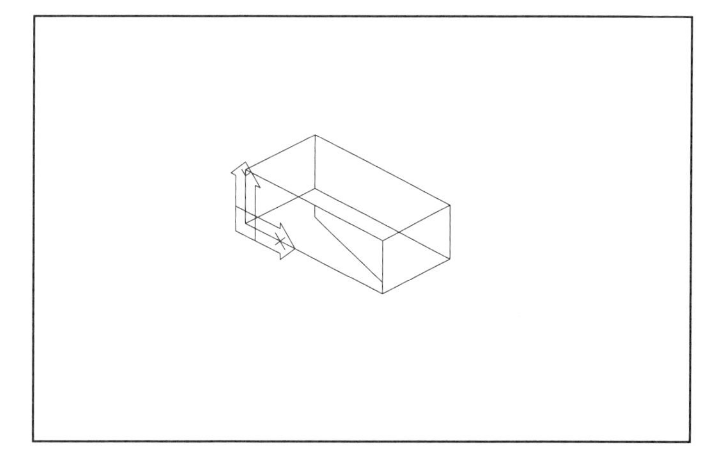

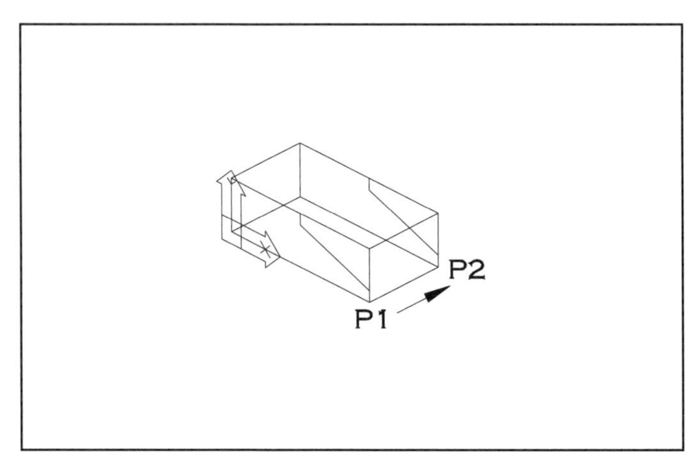

Figure 9.15.

We will use a multiple COPY to copy one of the front-to-back edges in three places.

☩ Enter the COPY command.
☩ Pick the bottom right edge for copying.
☩ Press Enter to end object selection.
☩ Type "m".
☩ Pick the front end point of the selected edge to serve as a base point of displacement.
☩ Pick the top end point of Line 1 (P1 in *Figure 9.16*).
☩ Pick the lower end point of Line 1 (P2) as another second point.
☩ Pick the right end point of Line 2 (P3) as another second point.
☩ Press Enter to exit the COPY command.

Finally, we need to do some trimming.

☩ Type "Tr" or select the Trim tool from the Modify toolbar.

For cutting edges, we want to select lines 1 and 2 and their copies in the back plane (lines 3 and 4 in *Figure 9.17*). Since this can be difficult, a quick alternative

Figure 9.16.

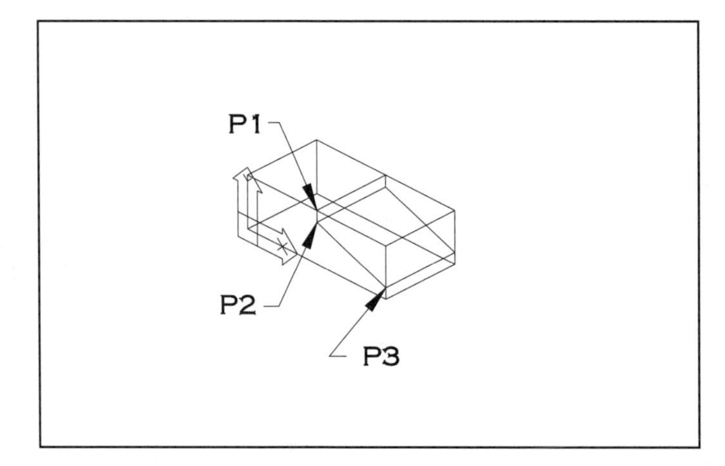

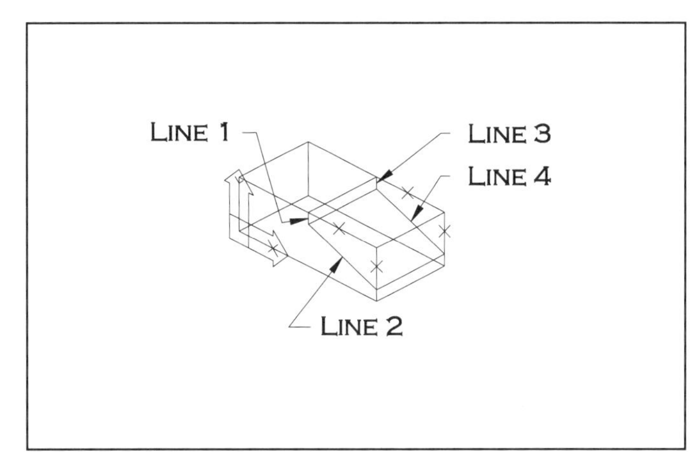

Figure 9.17.

to selecting these four separate lines is to use a crossing box to select the whole area. As long as your selection includes the four lines, it will be effective.

⌖ **Select the area around lines 1 and 2 with a crossing box.**

NOTE: Trimming in 3D can be tricky. Remember where you are. Edges that do not run parallel to the current UCS may not be recognized at all.

⌖ **Press Enter to end cutting edge selection.**

⌖ **One by one, pick the top front and top back edges to the right of the cut, and the right front and right back edges above the cut, as shown by the x's in *Figure 9.17*.**

⌖ **Press Enter to exit the TRIM command.**

⌖ **Enter the ERASE command and erase the top edge that is left hanging in space.** We use ERASE here because this line does not intersect any edges.

Your screen should now resemble *Figure 9.18.* This completes this wireframe figure. In the next section, we will use multiple tiled viewports and surface commands.

Figure 9.18.

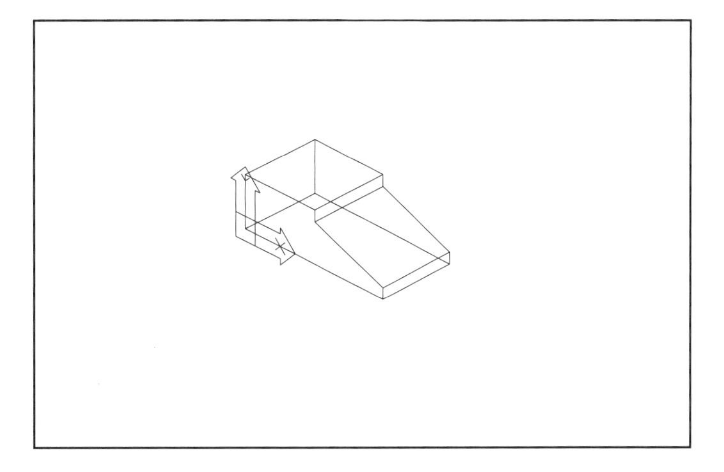

9.4 USING MULTIPLE TILED VIEWPORTS

A major feature needed to draw effectively in 3D is the ability to view an object from several different points of view simultaneously as you work on it. The VPORTS, or VIEWPORTS, command is easy to use and can save you from having to jump back and forth between different views of an object. Viewports can be used to place several 3D viewpoints on the screen at once. This can be a significant drawing aid. If you do not continually view an object from different points of view, it is easy to create entities that appear correct in the current view, but that are clearly incorrect from other points of view.

In this task we will divide your screen in half and define two views, so that you can visualize an object in plan view and a 3D view at the same time. As you work, remember that this is only a display command. The viewports we use in this chapter will be simple "tiled" viewports. Tiled viewports cover the complete drawing area, do not overlap, and cannot be plotted simultaneously. Plotting multiple viewports can be accomplished, but only in paper space with nontiled viewports.

⌗ Before going on, we will return to the World Coordinate System. It is always advisable to keep track of where you are in relation to the WCS and to begin there.

⌗ Type "ucs" or select "Set UCS" from the View pull down menu.

⌗ Type "w" or select "World" from the submenu.

Now we will move to two viewports.

⌗ Highlight "Tiled Viewports" on the View menu.

This will open the submenu shown in *Figure 9.19.* If you type the command you will see the same options at the command line prompt.

The "Layout . . ." option will call the Tiled Viewport Layout dialog box shown in *Figure 9.20.* Notice that the list on the left simply names the twelve layout options on the right. Any of these could also be specified by selecting a number of viewports from the submenu and then following the sequence of command prompts.

Figure 9.19.

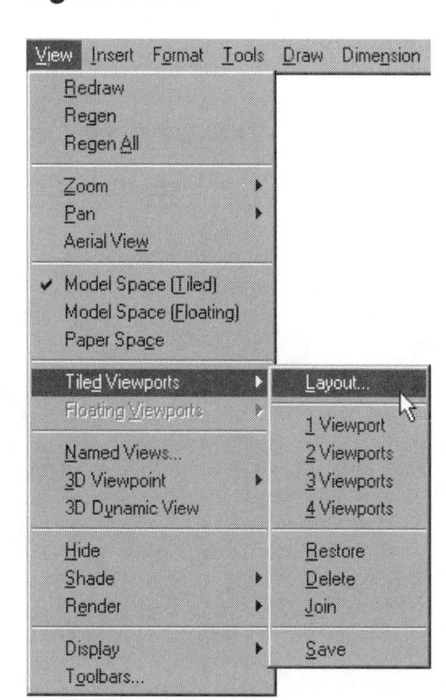

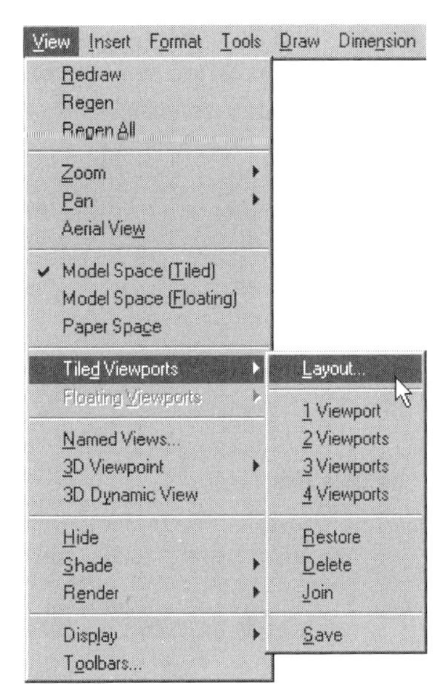

Figure 9.20.

For our purposes we will create a simple two-way vertical split, the "Two: Vertical" option in the dialog box.

⊕ **Select "Two: Vertical" or the box on the left end of the second row on the icon menu.**

⊕ **Click on "OK" to complete the dialog.**

Your screen will be regenerated to resemble *Figure 9.21.*

You will notice that the grid is rather small and confined to the lower part of the window and that the grid may be off in the left viewport. The shape of the viewports necessitates the reduction in drawing area. You can enlarge details as usual using the ZOOM command. Zooming and panning in one viewport will have no effect on other viewports. However, you can have only one UCS in effect at any time, so a change in the coordinate system in one viewport will be reflected in all viewports.

Figure 9.21.

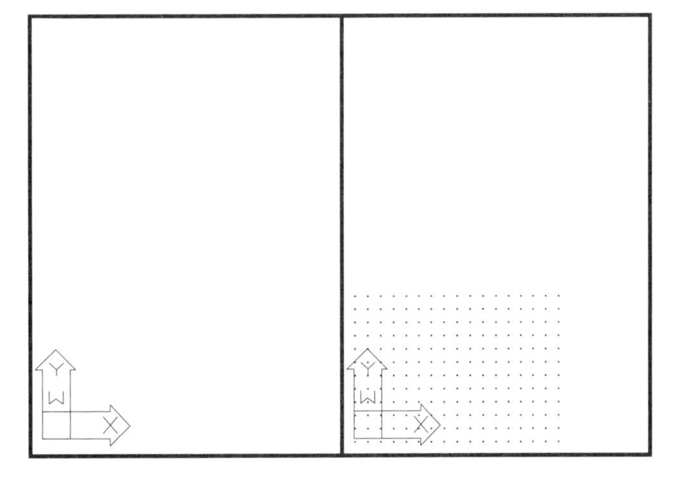

If you move your pointing device back and forth between the windows, you will see an arrow when you are on the left and the cross hairs when you are on the right. This indicates that the right window is currently active. Drawing and editing can be done only in the active window. To work in another window, you need to make it current by picking it with your pointing device. Often this can be done while a command is in progress.

⌗ **Move the cursor into the left window and press the pick button on your pointing device.**

Now the cross hairs will appear in the left viewport, and you will see the arrow when you move into the right viewport.

⌗ **If the grid is off in your left viewport, you may want to turn it on at this point.**

⌗ **Move the cursor back to the right and press the pick button again.**

This will make the right window active again.

There is no value in having two viewports if each is showing the same thing, so our next job will be to change the viewpoint in one of the windows. We will leave the window on the left in plan view and switch the right window to a plan view.

⌗ **Click in the left viewport to make it active.**

⌗ **Type "plan" or select "3D Viewpoint Presets," then "Plan View" from the pull down.**

⌗ **Type "w" or select "World" from the pull down submenu. In this case, "c" or "Current" will also work.**

Your screen should now be redrawn with a southeast isometric view in the right viewport as shown in *Figure 9.22.*

Once you have defined viewports, any drawing or editing done in the active viewport will appear in all the viewports. As you draw, watch what happens in both viewports.

Figure 9.22.

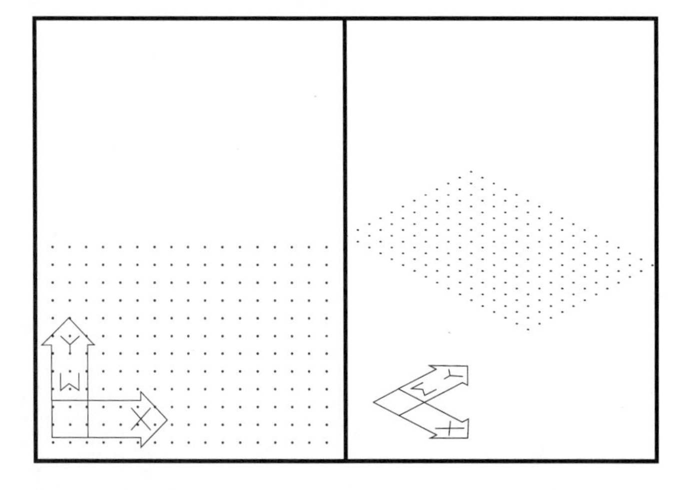

9.5 CREATING SURFACES WITH 3DFACE

3DFACE creates triangular and quadrilateral surfaces. 3D faces are built by entering points in groups of three or four to define the outlines of triangles or quadrilaterals, similar to objects formed by the 2D SOLID command. The surface of a 3D face is not shown on the screen, but it is recognized by the HIDE, SHADE, and RENDER commands.

In this exercise, we will add a surface to the top of the wedge block. You may eventually want a number of layers specifically defined for faces and surfaces, but this will not be necessary for the current exercise.

⌗ **Open the Draw menu and highlight Surfaces.**

This calls the submenu shown in *Figure 9.23.* There is also a Surface toolbar with a 3D face tool as shown in *Figure 9.24.*

⌗ **Select "3D Face".**

AutoCAD will prompt:

First point:

You can define points in either of the two viewports. In fact, you can even switch viewports in the middle of the command.

⌗ **Using an Endpoint object snap, pick a point similar to P1, as shown in *Figure* 9.25.**

Figure 9.23.

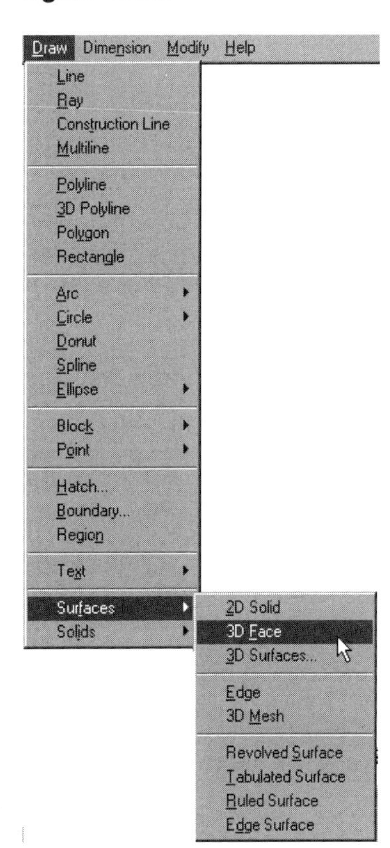

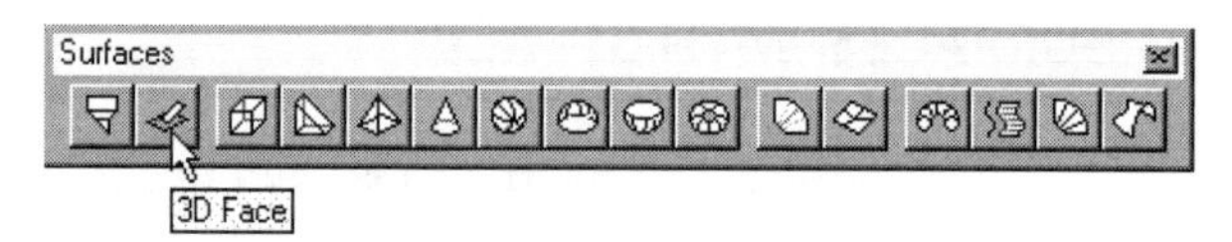

Figure 9.24.

AutoCAD prompts:

Second point:

⌖ **Using another Endpoint object snap, pick a second point, moving around the perimeter of the face, as shown.**
AutoCAD prompts:

Third point:

⌖ **Using an Endpoint object snap, pick a third point, as shown.**
AutoCAD prompts:

Fourth point:

NOTE: If you pressed Enter now, AutoCAD would draw the outline of a triangular face, using the three points already given.

⌖ **Using an Endpoint object snap, pick the fourth point of the face.**
AutoCAD draws the fourth edge of the face automatically when four points have been given, so it is not necessary to complete or close the rectangle.

AutoCAD will continue to prompt for third and fourth points so that you can draw a series of surfaces to cover an area with more than four edges. Keep in mind, however, that drawing faces in series is only a convenience. The result is a collection of independent three- and four-sided faces.

⌖ **Press Enter to exit the 3DFACE command.**

In the next sectioin, we will demonstrate the HIDE command.

Figure 9.25.

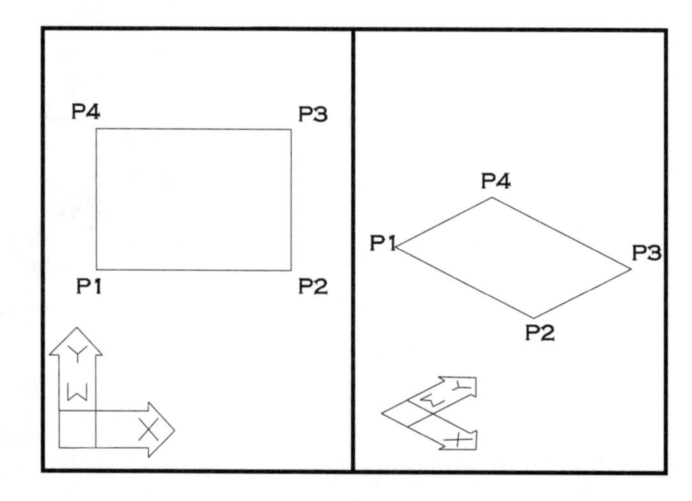

9.6 REMOVING HIDDEN LINES WITH HIDE

The HIDE command is easy to execute. However, execution may be slow in large drawings, and careful work may be required to create a drawing that hides the way you want it to. This is a primary objective of surface modeling. When you've got everything right, HIDE will temporarily remove all lines and objects that would be obstructed in the current view, resulting in a more realistic representation of the object in space. A correctly surfaced model can also be used to create a shaded rendering. Hiding has no effect on wireframe drawings, since there are no surfaces to obstruct lines behind them.

⌖ **Make the right viewport active.**

⌖ **Type "hi" or pen the View menu and select "Hide".**

Your screen should be regenerated to resemble *Figure 9.26.* Notice that the object has a "top" on it, the surface you created in the last task.

Figure 9.26.

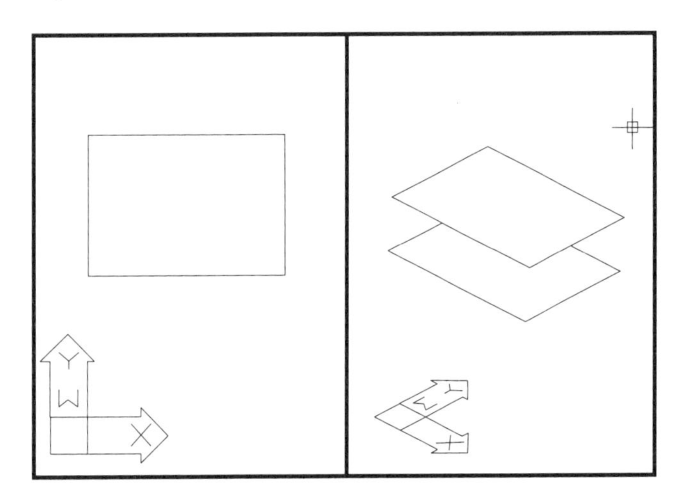

Following are some important points about hidden line removal which you should read before continuing:

1. Hidden line removal can be done from the plot configuration dialog box. However, due to the time involved and the difficulty of getting a hidden view just right, it is usually better to experiment on the screen first, then plot with hidden lines removed when you know you will get the image you want.
2. Whenever the screen is regenerated, hidden lines are returned to the screen.
3. Layer control is important in hidden line removal. Layers that are frozen are ignored by the HIDE command, but layers that are off are treated like layers that are visible. This can, for example, create peculiar blank spaces in your display if you have left surfaces or solids on a layer that is off at the time of hidden line removal.

9.7 USING 3D POLYGON MESH COMMANDS

3DFACE can be used to create simple surfaces. However, most surface models require large numbers of faces to approximate the surfaces of real objects. Consider the number of faces in *Figure 9.27.* Obviously you would not want to draw such an image one face at a time.

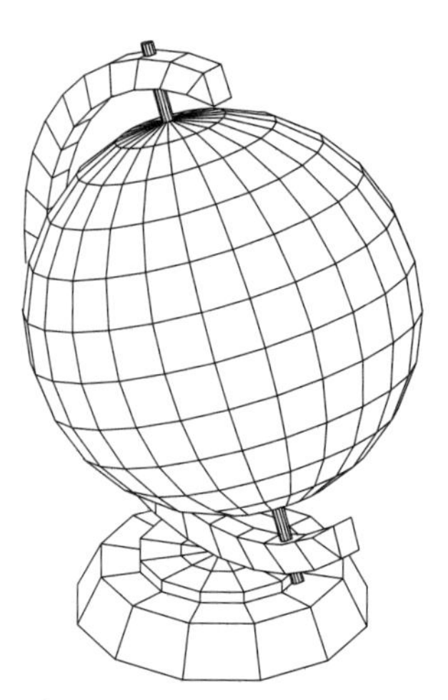

Figure 9.27.

AutoCAD includes a number of commands that make the creation of some types of surfaces very easy. These powerful commands create 3D polygon meshes. Polygon meshes are made up of 3D faces and are defined by a matrix of vertices.

⌗ **To begin this task, ERASE the wedge block from the previous tasks and draw an arc and a line below it as shown in *Figure 9.28. Exact sizes and locations are not important.***

The entities may be drawn in either viewport.

Now we will define some 3D surfaces using the arc and line you have just drawn.

TABSURF

The first surface we will draw is called a "tabulated surface." In order to use the TABSURF command, you need a line or curve to define the shape of the surface and a vector to show its size and direction. The result is a surface generated by repeating the shape of the original curve at every point along the path specified by the vector.

Figure 9.28.

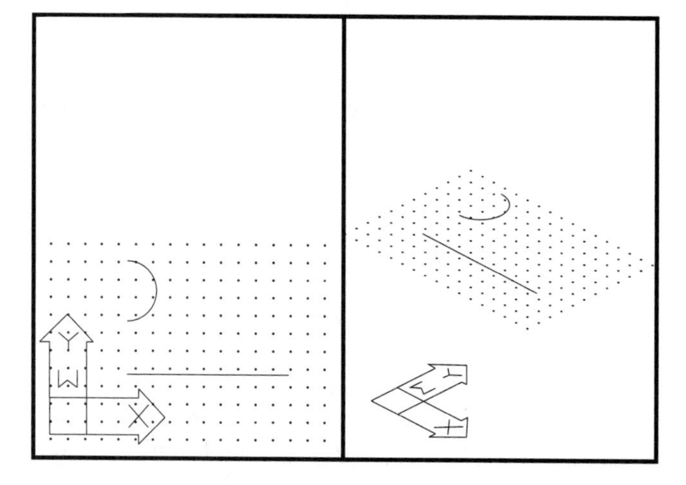

Type "Tabsurf" or open the Draw menu, highlight Surfaces and select Tabulated Surface.

There is also a Surface toolbar with a Tabulated Surface tool as shown previoulsy in *Figure 9.24.* AutoCAD will prompt:

Select path curve:

The path curve is the line or curve that will determine the shape of the surface. In our case it will be the arc.

Pick the arc.

AutoCAD will prompt for a vector:

Select direction vector:

We will use the line. Notice that the vector does not need to be connected to the path curve. Its location is not significant, only its direction and length.

There is an oddity here to watch out for as you pick the vector. If you pick a point near the left end of the line, AutoCAD will interpret the vector as extending from left to right. Accordingly, the surface will be drawn to the right. By the same token, if your point is near the right end of the line, the surface will be drawn to the left. Most of the time you will avoid confusion by picking a point on the side of the vector nearest the curve itself.

Pick a point on the left side of the line.

Your screen will be redrawn to resemble *Figure 9.29.*

Notice that this is a flat surface even though it may look 3D. If this is not obvious in the plan view, it should be more apparent in the 3D view.

Tabulated surfaces can be fully 3D, depending on the path and vector chosen to define them. In this case we have an arc and a vector that are both entirely in the XY plane, so the resulting surface is also in that plane.

RULESURF

TABSURF is useful in defining surfaces that are the same on both ends, assuming you have one end and a vector. Often, however, you have no vector, or you need to draw a surface between two different paths. In these cases you will need the RULESURF command.

Figure 9.29.

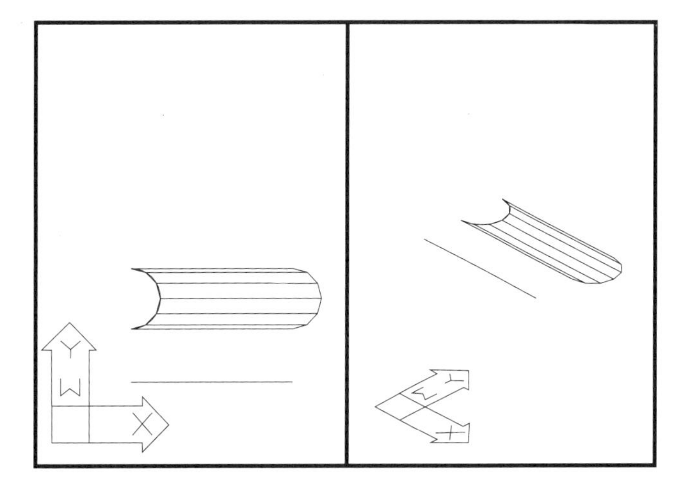

For example, what if we need to define a surface between the line and the arc? Let's try it.

- **Type "u" to undo the last tabulated surface.**
- **Open the Draw menu, highlight Surfaces, and select Ruled Surface.**

The first prompt is:

Select first defining curve:

- **Pick the arc, using a point near the bottom.**

Remember that you must pick points on corresponding sides of the two defining curves in order to avoid an hourglass effect.

AutoCAD prompts:

Select second defining curve:

- **Pick the line, using a point near the left end.**

Your screen should resemble *Figure 9.30.*

Again, notice that this surface is within the XY plane even though it may look 3D. Ruled surfaces may be drawn just as easily between curves that are not coplanar.

EDGESURF

TABSURF creates surfaces that are the same at both ends and move along a straight line vector. RULESURF draws surfaces between any two boundaries. There is a third command, EDGESURF, which draws surfaces that are bounded by four curves. Edge-defined surfaces have a lot of geometric flexibility. The only restriction is that they must be bounded on all four sides. That is, they must have four edges that touch.

In order to create an EDGESURF, we need to undo our last ruled surface and add two more edges.

- **Type "u".**
- **Add a line and an arc to your screen, as shown in Figure 9.31.**

Remember, you can draw in either viewport.

- **Open the Draw menu, highlight Surfaces, and select Edge Surface.**

Figure 9.30.

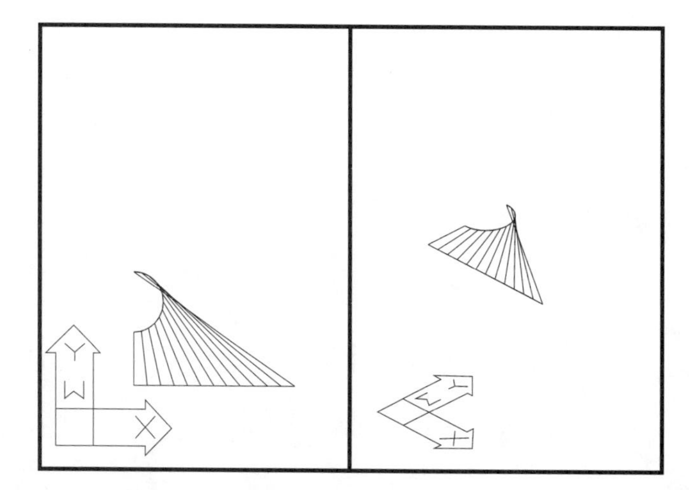

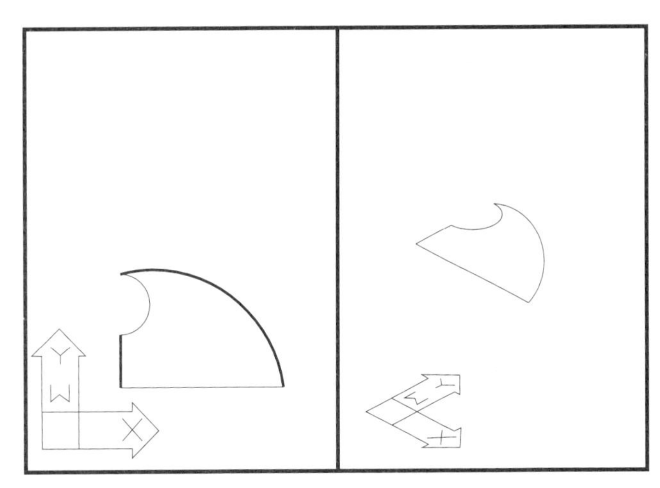

Figure 9.31.

AutoCAD will prompt for the four edges of the surface, one at a time:

Select edge 1:

⌖ **Pick the smaller arc.**
AutoCAD prompts:

Select edge 2:

⌖ **Pick the larger arc.**
AutoCAD prompts:

Select edge 3:

⌖ **Pick the longer line.**
AutoCAD prompts:

Select edge 4:

⌖ **Pick the smaller line.**
Your screen should now resemble *Figure 9.32.*

Figure 9.32.

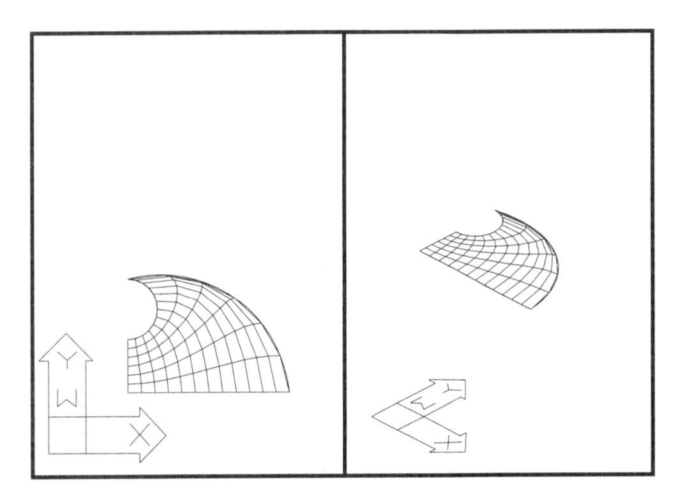

REVSURF

We have one more 3D polygon mesh command to explore, and this one is probably the most impressive of all. REVSURF creates surfaces by spinning a curve through a given angle around an axis of revolution. Just as tabulated surfaces are spread along a linear path, surfaces of revolution follow a circular or arc-shaped path. As a result, surfaces of revolution are always fully three-dimensional, even if their defining geometry is in a single plane, as it will be here.

⌖ **In preparation for this exercise undo the EDGESURF, so that your screen resembles *Figure 9.31* again.**

We will use REVSURF to create a surface of revolution.

⌖ **Open the Draw menu, highlight Surfaces and select Revolved Surface.**

AutoCAD needs a path curve and an axis of revolution to define the surface. The first prompt is:

Select path curve:

⌖ **Pick the smaller arc.**

AutoCAD prompts:

Select axis of revolution:

⌖ **Pick the smaller line.**

AutoCAD now needs to know whether you want the surface to begin at the curve itself or somewhere else around the circle of revolution:

Start angle <0>:

The default is to start at the curve.

⌖ **Press Enter to begin the surface at the curve itself.**

AutoCAD prompts:

Included angle (+=ccw, −=cw) <Full circle>:

Entering a positive or negative degree measure will cause the surface to be drawn around an arc rather than a full circle. The default will give us a complete circle.

⌖ **Press Enter.**

Your screen should be drawn to resemble *Figure 9.33*.

Figure 9.33.

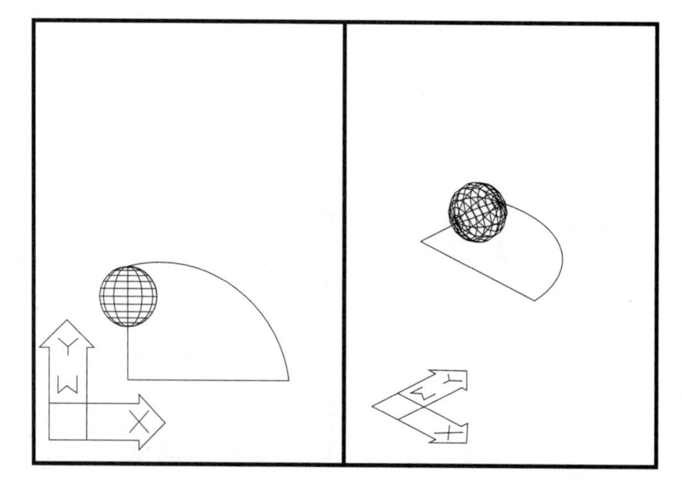

Next, we will move on to solid modeling, probably the most powerful type of 3D drawing.

9.8 CREATING SOLID BOXES AND WEDGES

Solid modeling requires a somewhat different type of thinking from any of the drawings you have done so far. Instead of focusing on lines and arcs, edges and surfaces, you will need to imagine how 3D objects might be pieced together by combining or subtracting basic solid shapes. This building block process is called constructive solid geometry and includes joining, subtracting, and intersecting operations. A simple washer, for example, could be made by cutting a small cylinder out of the middle of a larger cylinder. In AutoCAD solid modeling you can begin with a flat outer cylinder, then draw an inner cylinder with a smaller radius centered at the same point, and then subtract the inner cylinder from the outer, as illustrated in *Figure 9.34.*

This operation, which uses the SUBTRACT command, is the equivalent of cutting a hole, and is one of three Boolean operations (after the mathematician George Booles) used to create composite solids. UNION joins two solids to make a new solid, and INTERSECT creates a composite solid in the space where two solids overlap.

In this exercise, you will create a composite solid from the union and subtraction of several solid primitives. Primitives are 3D solid building blocks-boxes, cones, cylinders, spheres, wedges, and torus. They all are regularly shaped and can be defined by specifying a few points and distances.

For this exercise, we will return to a single 3D viewpoint.

☩ **Erase objects left from the last task.**
☩ **Make the right viewport active.**

Figure 9.34.

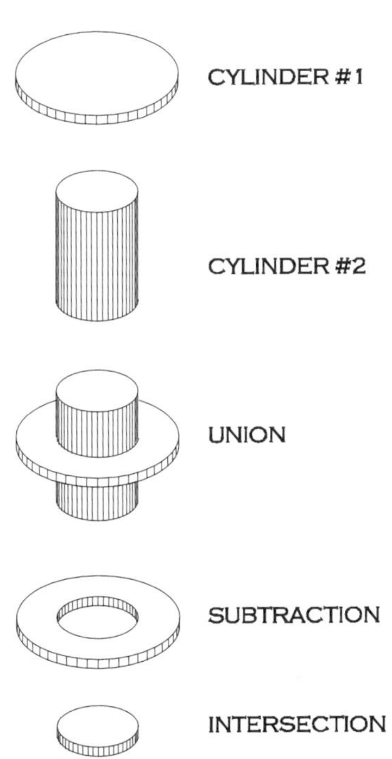

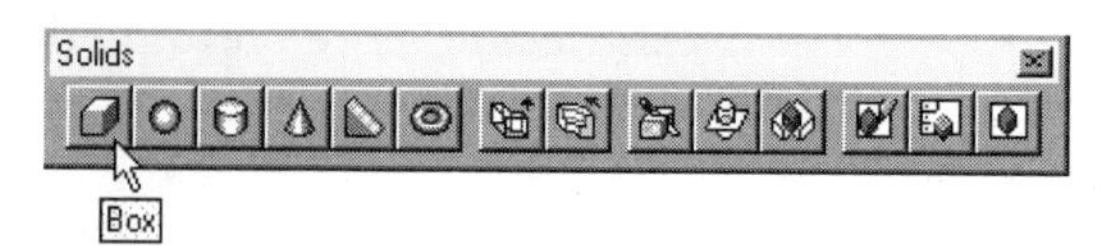

Figure 9.35.

⌖ Select Tiled Viewports from the View pull down menu
⌖ From the submenu, select "1 Viewport".
We are now ready to create new solid objects.
⌖ Type "box" or open the Draw menu, highlight Solids, and select Box.
There is also a Solids toolbar with a Box tool, as illustrated in *Figure 9.35.*
In the command area you will see the following prompt:

Center/<Corner of box> <0,0,0>:

"Center" allows you to begin defining a box by specifying its center point. Here we will use the "Corner of box" option to begin drawing a box in the baseplane of the current UCS. We will draw a box with a length of 4, width of 3, and height of 1.5.

⌖ Pick a corner point similar to point 1 in *Figure 9.36.*
AutoCAD will prompt:

Cube/Length/<Other corner>:

With the "Cube" option you can draw a box with equal length, width, and height simply by specifying one distance. The "Length" option will allow you to specify length, width, and height separately. If you have simple measurements that fall on snap points, as we do, you can show the length and width at the same time by picking the other corner of the base of the box (the default method).

⌖ Move the cross hairs over 4 in the x direction and up 3 in the y direction to point 2, as shown in the figure.

Notice that "length" is measured along the x axis, and "width" is measured along the y axis.

Figure 9.36.

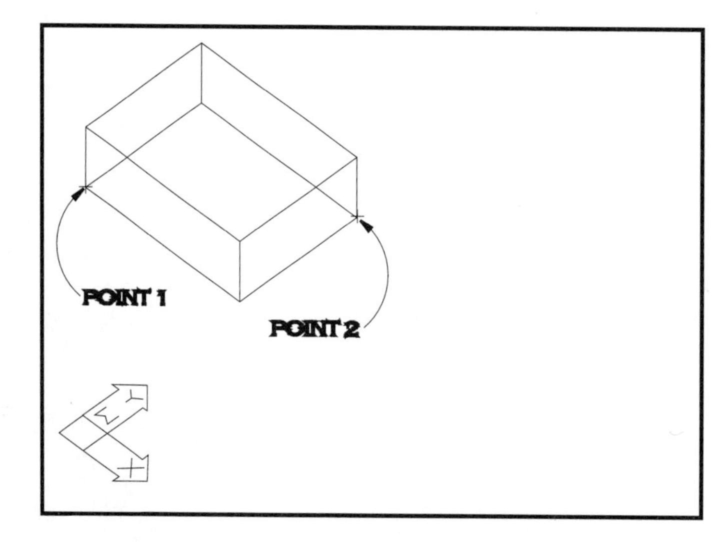

Pick point 2 as shown.

Now AutoCAD prompts for a height. "Height" is measured along the z axis. As usual, you cannot pick points in the z direction unless you have objects to snap to. Instead, you can type a value or show a value by picking two points in the XY plane.

Type "1.5" or pick two points 1.5 units apart.

Your screen should resemble *Figure 9.36.*

Next we will create a solid wedge. The process will be exactly the same, but there will be no "Cube" option.

Type "we" or open the Draw menu, highlight Solids and select Wedge.

AutoCAD prompts:

Center/<Corner of wedge> <0,0,0>:

Pick the front corner point of the box, as shown in *Figure* 9.37.

As in the BOX command, AutoCAD prompts for a cube, length, or the other corner:

Cube/Length/<other corner>:

This time let's use the length and width option.

Type "L".

The rubber band will disappear and AutoCAD will prompt for a length, which you can define by typing a number or showing two points.

Type "4" or show a length of 4.00 units.

AutoCAD now prompts for a width. Remember, length is measured in the x direction, and width is measured in the y direction.

Type "3" or show a width of 3 units.

AutoCAD prompts for a height.

Type "3" or show a distance of 3 units.

AutoCAD will draw the wedge you have specified. Notice that a wedge is simply half a box, cut along a diagonal plane.

Figure 9.37.

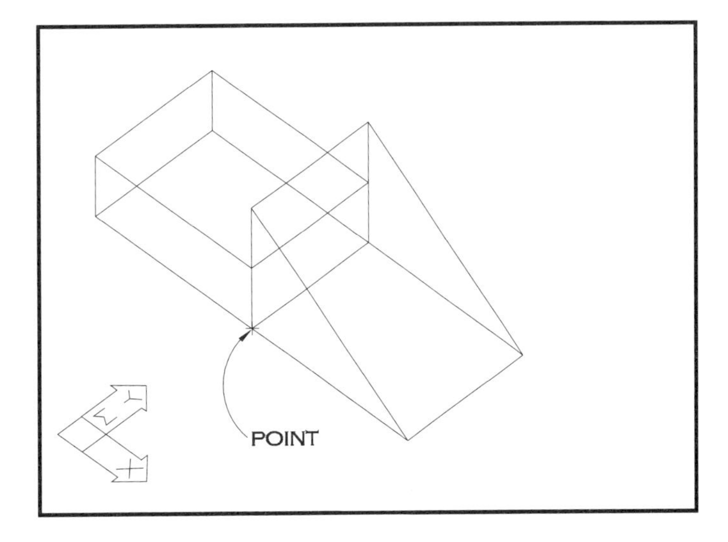

Your screen should resemble *Figure 9.37*. Although the box and the wedge appear as wireframe objects, they are really quite different, as you will find. In Task 2 we will join the box and the wedge to form a new composite solid.

9.9 CREATING THE UNION OF TWO SOLIDS

Unions are simple to create and usually easy to visualize. The union of two objects is an object that includes all points that are on either of the objects. Unions can be performed just as easily on more than two objects. The union of objects can be created even if the objects have no points in common (i.e., they do not touch or overlap).

Although the box and wedge are adjacent, there are still two distinct solids on the screen; with UNION we can join them.

⌖ **Type "Union" or open the Modify menu, highlight Boolean, and select Union.**

There is also a Union tool on the Modify II toolbar, as shown in *Figure 9.38*.

Figure 9.38.

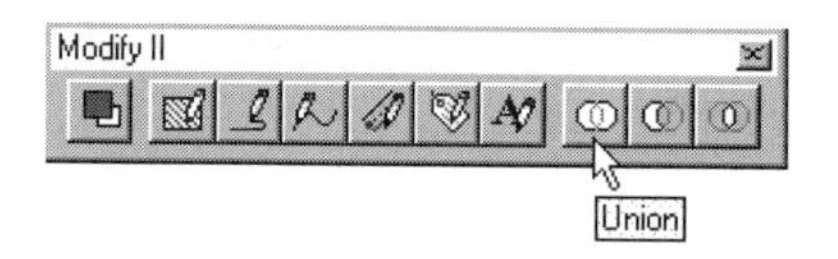

AutoCAD will prompt you to select objects.

⌖ **Point or use a crossing box to select both objects.**

⌖ **Press Enter to end object selection.**

Your screen should resemble *Figure 9.39*.

Figure 9.39.

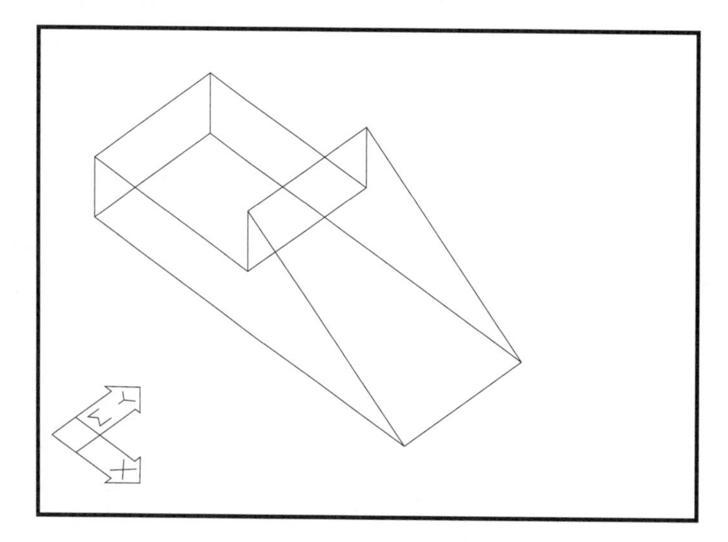

9.10 WORKING ABOVE THE XY PLANE USING ELEVATION

In this section we will draw two more solid boxes while demonstrating the use of elevation to position objects above the XY plane of the current UCS. Changing elevations simply adds a single Z value to all new objects as they are drawn and can be used as an alternative to creating a new UCS. With an elevation of 1.00, for example, new objects would be drawn 1.00 above the XY plane of the current UCS. You can also use a thickness setting to create 3D objects, but these will be created as surface models, not solids.

We will begin by drawing a second box positioned on top of the first box. Later we will move it, copy it, and subtract it to form a slot in the composite object.

⌗ **Type "Elev".**

AutoCAD prompts:

New current elevation <0.00>:

The elevation is always set at 0 unless you specify otherwise.

⌗ **Type "1.5"**

This will bring the elevation up 1.5 out of the XY plane, putting it even with the top of the first box you drew.

AutoCAD now prompts for a new current thickness. Thickness does not apply to solid objects because they have their own thickness.

⌗ **Press Enter to retain 0.00 thickness.**

This brings us back to the command prompt. If you watch closely you will see that the grid has also moved up into the new plane of elevation.

⌗ **Type "Box" or open the Draw menu, highlight Solids, and select Box.**

⌗ **Pick the upper left corner of the box, point 1 in *Figure 9.40.***

⌗ **Type "L" to initiate the "length" option.**

⌗ **Type "4" or pick two points (points 1 and 2 in the figure) to show a length of 4 units.**

⌗ **Type ".5" or pick two points (points 1 and 3) to show a width of 0.5 units.**

Figure 9.40.

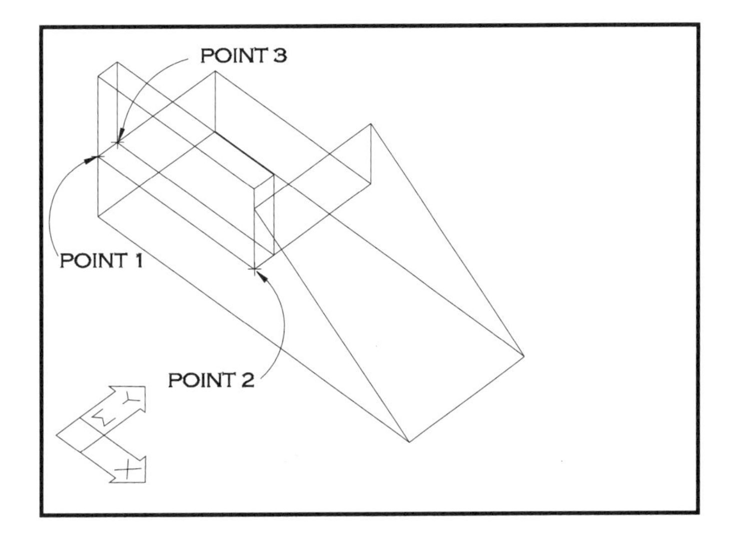

⌗ **Type "2" or pick two points to show a height of 2 units.**

Your screen should resemble *Figure 9.40.* Notice how you were able to pick points on top of the box because of the change in elevation just as if we had changed coordinate systems. Before going on, return to 0.00 elevation.

⌗ **Type "elev".**

⌗ **Type "0".**

⌗ **Press Enter to retain 0.00 thickness.**

9.11 CREATING COMPOSITE SOLIDS WITH SUBTRACT

SUBTRACT is the logical opposite of UNION. In a union operation, all the points contained in one solid are added to the points contained in other solids to form a new composite solid. In a subtraction, all points in the solids to be subtracted are removed from the source solid. A new composite solid is defined by what is left.

In this exercise, we will use the objects already on your screen to create a slotted wedge. First we need to move the thin upper box into place, then we will copy it to create a longer slot, and finally we will subtract it from the union of the box and wedge.

Before subtracting, we will move the box to the position shown in *Figure 9.41.*

⌗ **Type "m" or select the Move tool from the Modify toolbar.**

⌗ **Select the narrow box drawn in the last task.**

⌗ **Press Enter to end object selection.**

⌗ **At the "Base point or displacement" prompt use a midpoint object snap to pick the midpoint of the top right edge of the narrow box.**

⌗ **At the "Second point of displacement" prompt use another midpoint object snap to pick the top edge of the wedge, as shown in *Figure 9.41.***

This will move the narrow box over and down. If you were to perform the subtraction now, you would create a slot but it would only run through the box,

Figure 9.41.

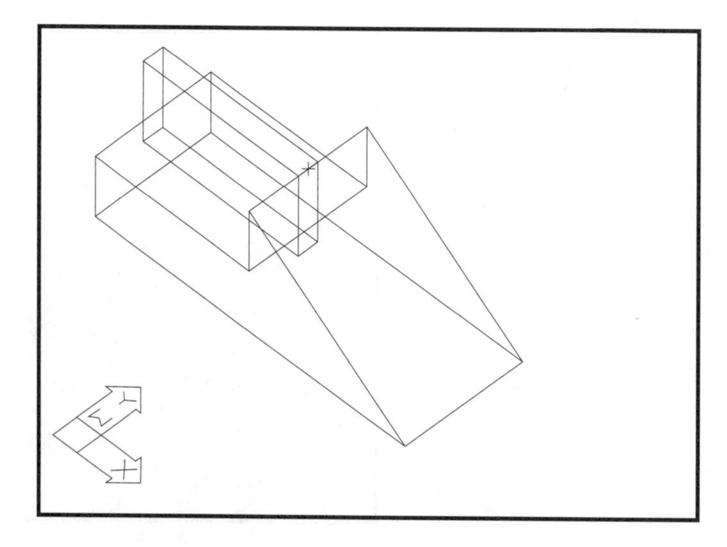

not the wedge. We can create a longer slot by copying the narrow box over to the right using grips.

Select the narrow box.

Pick any of the eight grips.

If ortho is off, turn it on.

Now if you move the cursor in the x direction you will see a copy of the box moving with you. Solids cannot be stretched, so we must make a copy to lengthen the slot.

Type "c".

Move the cursor between 2.00 and 4.00 units to the right and press the pick button.

If you don't go far enough the slot will be too short. If you go past 4.00, the slot will be interrupted by the space between the box and the copy.

Press Enter to leave the grip edit mode.

Your screen will resemble *Figure 9.42.*

Type "su" or open the Modify menu, highlight Boolean and select Subtract.

There is also a Subtract tool on the Modify II toolbar as shown previously in *Figure 9.39.*

AutoCAD asks you to select objects to subtract from first:

    Select solids and regions to subtract from . . .
    Select objects:

Pick the composite of the box and the wedge.

Press Enter to end selection of source objects.

AutoCAD prompts for objects to be subtracted:

    Select solids and regions to subtract . . .
    Select objects:

Pick the two narrow boxes.

Press Enter to end selection.

Your screen will resemble *Figure 9.43.*

Figure 9.42.

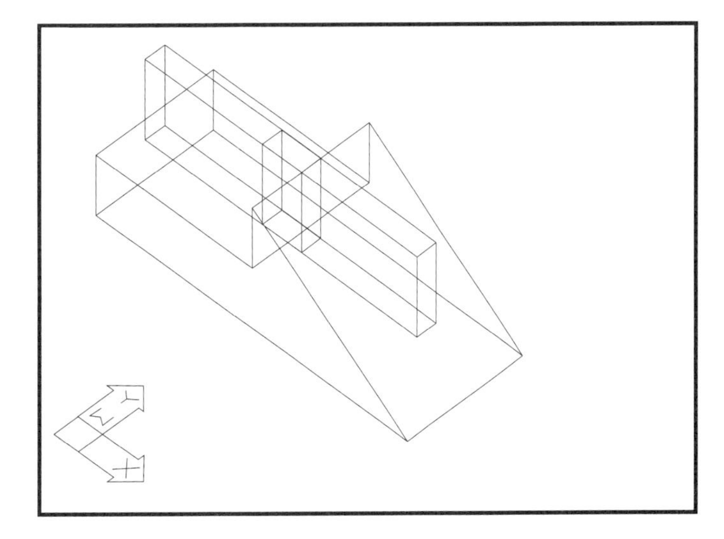

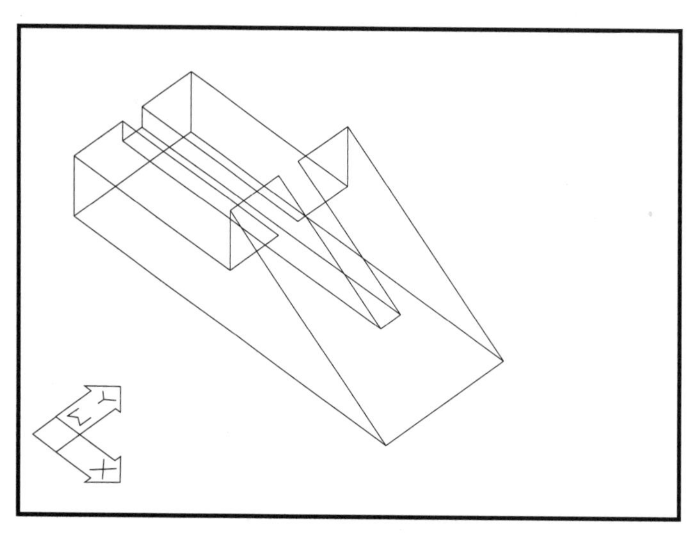

Figure 9.43.

9.12 SHADING AND RENDERING

There are many ways to present a solid object. Once drawn you can view it from any angle, create a perspective view using the DVIEW command, remove hidden lines, or create dramatic shaded or rendered images. Before leaving, try these three simple procedures. We hope that these will whet your appetite and that you will go on to learn much more about 3D imaging in AutoCAD.

⌖ **To remove hidden lines, type "hide".**

This will produce a simple hidden line image that clearly shows how AutoCAD is recognizing the surfaces of this solid object.

⌖ **To create a simple shaded imade, type "shade".**

This will produce a shaded image similar to *Figure 9.44.*

⌖ **To create a simple rendering, type "render". This will call up the Render dialogue box.**

⌖ **In the Render dialogue box, select "Render Scene" in the lower left corner.**

This will produce a simple rendered image similar to *Figure 9.45.* In a rendered image you can create and control effects to simulate light, shadow, and different

Figure 9.44.

Figure 9.45.

materials. Although lighting is beyond the scope of this introduction to AutoCAD, we would encourage you to explore the LIGHT command, which allows you to add three different types of lighting to rendered images, and the RMAT and MATLIB commands, which allow you to create the effect of different materials. Good luck!

9.13 REVIEW MATERIAL

Questions

1. What is a wireframe model?
2. What is the significance of the W on the UCS icon?
3. What coordinates indicate a displacement of –5 in the z direction from the point (6,6,6)?
4. Why is it usually necessary to utilize object snap to select a point on an object outside of the XY plane of the current UCS?
5. What information defines a user coordinate system?
6. Why is it important to keep two or more different views of an object on the screen as you are drawing and editing it?
7. What two basic geometric shapes can be drawn as a 3DFACE?
8. How does the HIDE command treat objects on layers that are turned off?
9. What geometry is needed to define a tabulated surface? A ruled surface? An edge surface? A revolved surface?
10. What 3D solid objects and commands would you use to create a square nut with a bolt hole in the middle?
11. What command allows you to draw outside of the xy plane of the current UCS without the use of typed coordinates, object snap, or point filters?

COMMANDS

View	Surfaces	Solids
UCS	3DFACE	BOX
UNION	EDGESURF	WEDGE
DDUCS	REVSURF	
UCSICON	RULESURF	
	TABSURF	

Drawing Problems

1. Setup a southeast isometric 3D viewpoint in the world coordinate system and draw a regular hexagon with a circumscribed radius of 4.0 units.
2. Create a half sized scaled copy of the hexagon centered at the same center as the original hexagon, then move the smaller hexagon 5.0 units up in the z direction.
3. Connect corresponding corners of the two hexagons to create tapered hexagonal prism in 3 dimensions.
4. Create a ucs aligned with any of the faces of the hexagonal prism.

PROFESSIONAL SUCCESS

Using Solid Models

Techniques have been developed that allow solid models to be used for purposes besides visualization. The solid model, which contains a complete mathematical representation of the object's surface and interior, is easily converted into specialized computer code. Such software is used in applications that include stress analysis and computer-aided manufacturing.

A finite element analysis (FEA) allows the deformation behavior of an object under mechanical or thermal loads to be studied in detail. The solid model is used by an FEA preprocessor program to sub-divide the object under study into a three-dimensional mesh consisting of simple rectangular or triangular elements. The simple geometry of the finite elements allows a system of equations to be generated by the FEA software, which, when solved using the applied loads, provides stress, displacement (strain), and thermal maps of the solid. These maps are often displayed as color contour plots superimposed on the solid model, and allow areas of high stress and strain to be visualized.

Solid models may also be used to directly generate manufactured prototypes without any human intervention. Computer numerical control (CNC) machine tools are controlled with programs automatically generated from the solid model. Such programs control the trajectory of the tool as it travels over a part, causing the desired surfaces to be cut with high precision. CNC tools remove the surface variability associated with manual machine tools controlled by human operators, and allow many identical parts to be machined automatically.

DRAWING 9–1

Clamp

This drawing is similar to the one you did in the chapter. Two major differences are that it is drawn from a different viewpoint and that it includes dimensions in the 3D view. This will give you additional practice in defining and using User Coordinate Systems. Your drawing should include dimensions, border, and title.

Drawing Suggestions

- We drew the outline of the clamp in a horizontal position and then worked from a northeast isometric or back, left, top point of view.
- Begin in WCS plan view drawing the horseshoe-shaped outline of the clamp. This will include fillets on the inside and outside of the clamp. The more you can do in plan view before copying to the top plane, the less duplicate editing you will need to do later.
- When the outline is drawn, switch to a northeast isometric view.
- COPY the clamp outline up 1.50.
- Define User Coordinate Systems as needed, and save them whenever you are ready to switch to another UCS. You will need to use them in your dimensioning.
- The angled face, the slots, and the filleted surfaces can be drawn just as in the chapter.

Dimensioning in 3D

The trick to dimensioning a 3D object is that you will need to restore the appropriate UCS for each set of dimensions. Think about how you want the text to appear. If text is to be aligned with the top of the clamp (i.e., the 5.75 overall length), you will need to draw that dimension in a "top" UCS; if it is to align with the front of the object (the 17 degree angle and the 1.50 height), draw it in a "front" UCS, and so forth.

- Define a UCS with the "View" option in order to add the border and title. Type "UCS", then "v". This creates a UCS aligned with your current viewing angle.

Setting Surftab1

Notice that there are 16 lines defining the RULESURF fillets in this drawing, compared to 6 in the chapter. This is controlled by the setting of the variable Surftab1. You can change it by typing "Surftab1" and entering "16" for the new value.

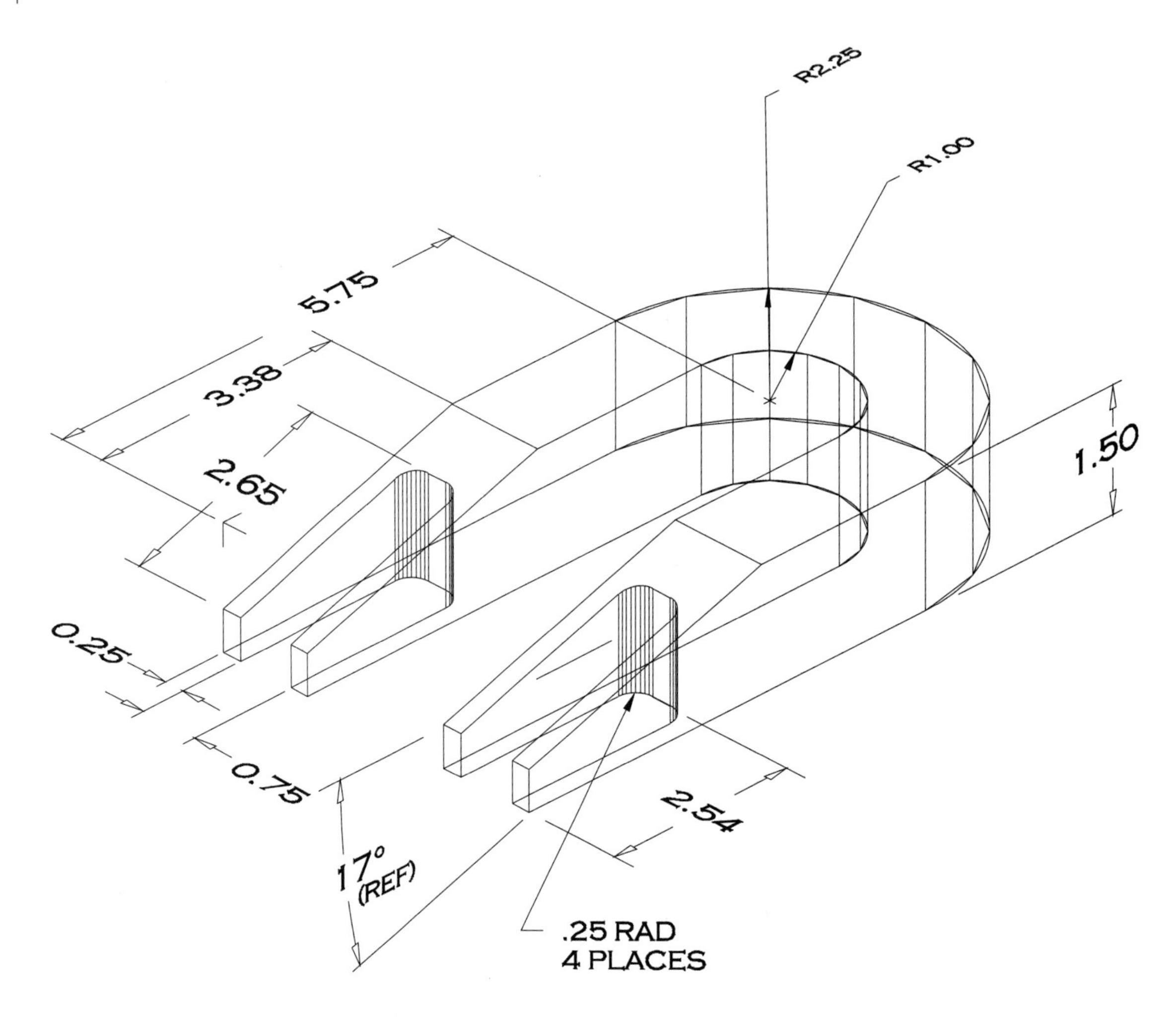

CLAMP

DRAWING 9-1

DRAWING 9–2

Revsurf Designs

The REVSURF command is fascinating and powerful. As you get familiar with it, you may find yourself identifying objects in the world that can be conceived as surfaces of revolution. To encourage this process, we have provided this page of 12 REVSURF objects and designs.

To complete the exercise, you will need only the PLINE and REVSURF commands. In the first six designs we have shown the path curves and axes of rotation used to create the design. In the other six you will be on your own.

Exact shapes and dimensions are not important in this exercise. Imagination is. When you have completed our designs, we encourage you to invent a number of your own.

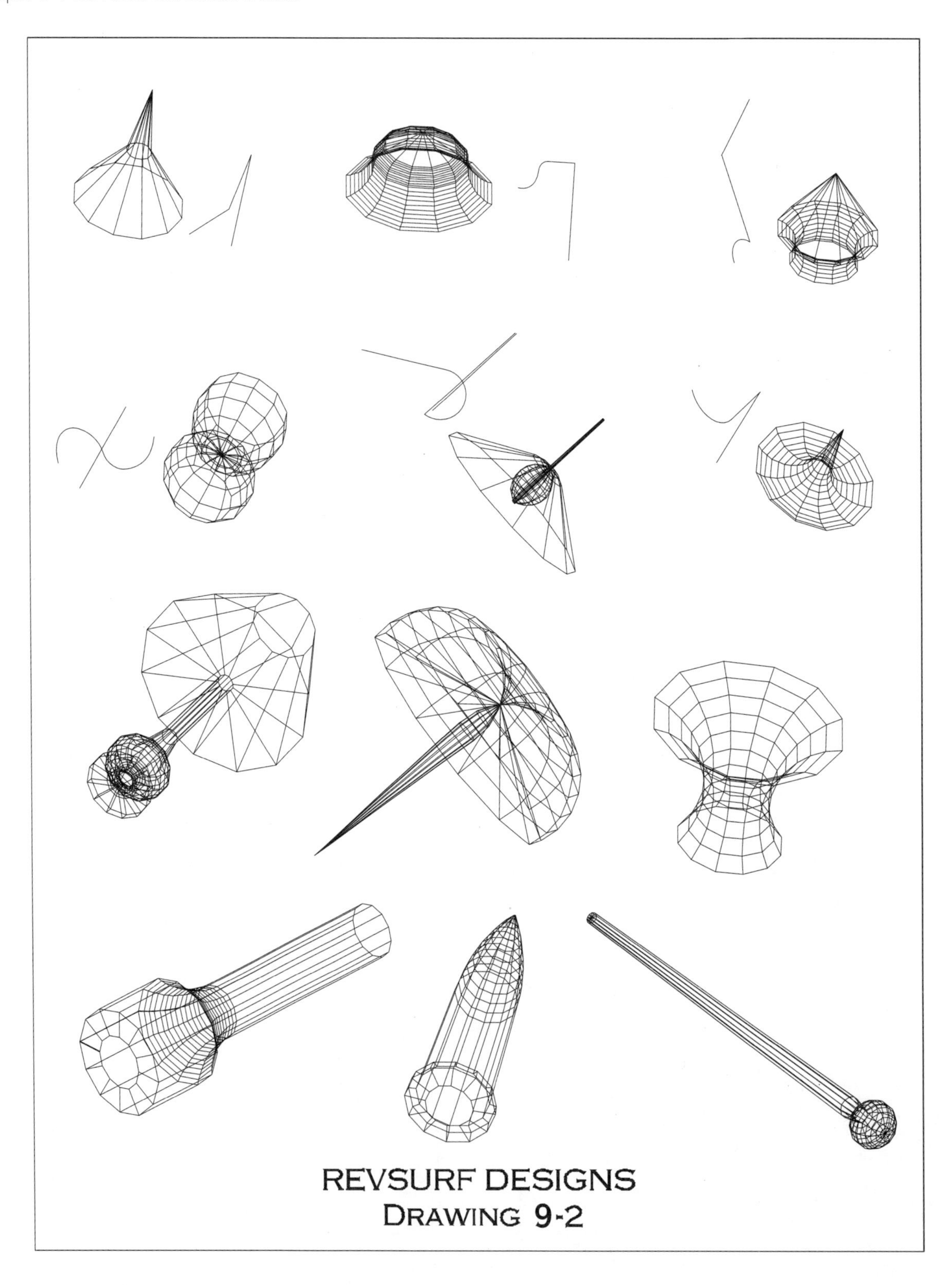
REVSURF DESIGNS
DRAWING 9-2

DRAWING 9-3

Bushing Mount

We offer the drawing suggestions following for Drawing 9-3. The principles demonstrated here will carry over into the other drawings, and you should be capable of handling them on your own at this point. In all cases you are encouraged to experiment with renderings when the model is complete.

Use an efficient sequence in the construction of composite solids. In general this will mean saving union, subtraction, and intersection operations until most of the solid objects have been drawn and positioned. This approach allows you to continue to use the geometry of the parts for snap points as you position other parts.

Drawing Suggestions

- Use at least two views, one plan and one 3D, as you work.
- Begin with the bottom of the mount in the XY plane. This will mean drawing a 6.00 × 4.00 × .50 solid box sitting on the XY plane.
- Draw a second box, 1.50 × 4.00 × 3.50, in the XY plane. This will become the upright section at the middle of the mount. Move it so that its own midpoint is at the midpoint of the base.
- Draw a third box, 1.75 × .75 × .50, in the XY plane. This will be copied and become one of the two slots in the base. Move it so that the midpoint of its long side is at the midpoint of the short side of the base. Then MOVE it 1.125 along the x axis.
- Add a .375 radius cylinder with .50 height at each end of the slot.
- Copy the box and cylinders 3.75 to the other side of the base to form the other slot.
- Create a new UCS 2.00 up in the z direction. You can use the origin option and give (0,0,2) as the new origin. This puts the XY plane of the UCS directly at the middle of the upright block, where you can easily draw the bushing.
- Move out to the right of the mount and draw the polyline outline of the bushing as shown in the drawing. Use REVOLVE to create the solid bushing.
- Create a cylinder in the center of the mount, where it can be subtracted to create the hole in the mount upright.
- Union the first and second boxes.
- Subtract the boxes and cylinders to form the slots in the base and the bushing-sized cylinder to form the hole in the mount.

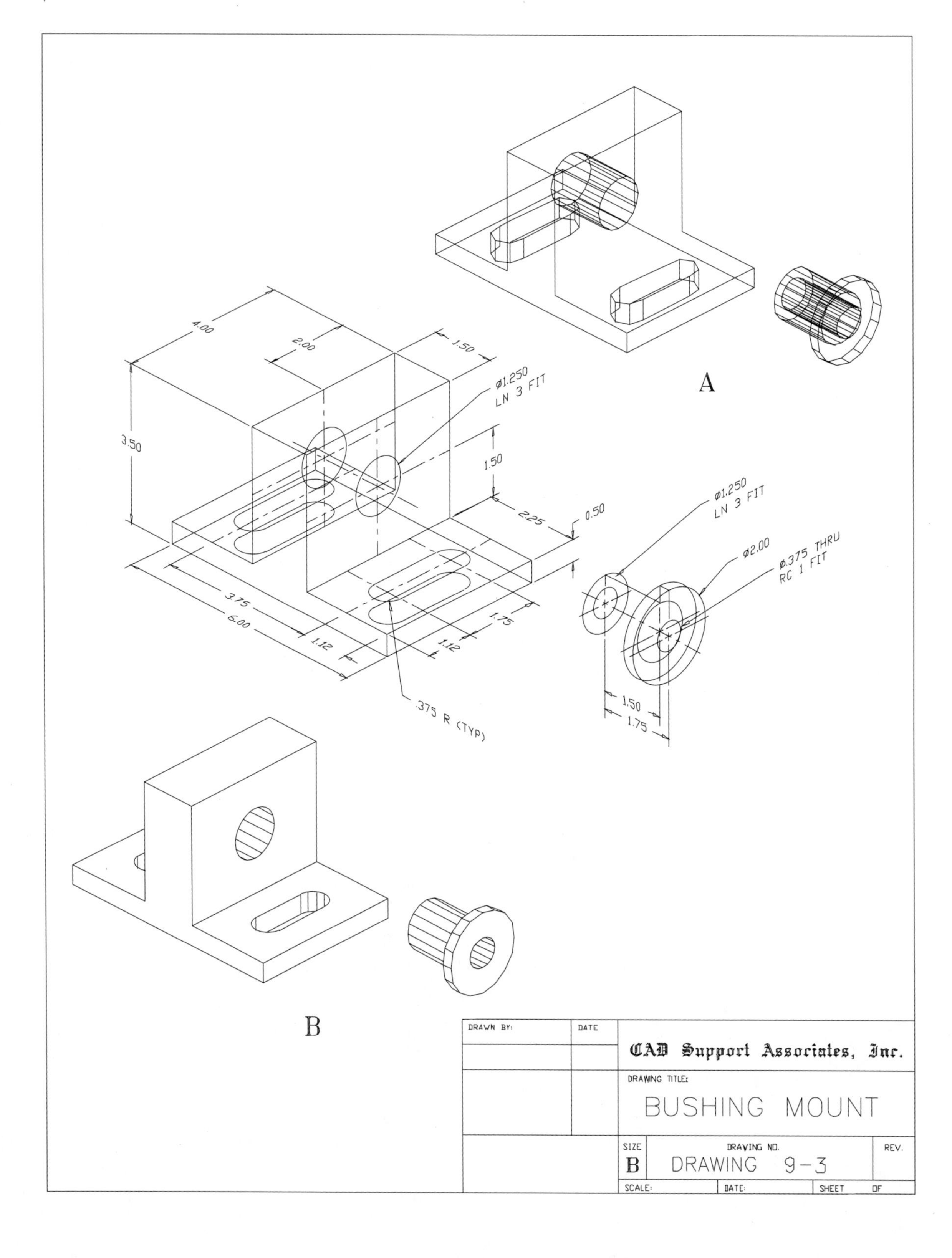
A
4.00
2.00
1.50
Ø1.250
LN 3 FIT
3.50
1.50
2.25
0.50
Ø1.250
LN 3 FIT
Ø2.00
Ø.375 THRU
RC 1 FIT
3.75
6.00
1.12
1.75
1.12
.375 R (TYP)
1.50
1.75
B
DRAWN BY:
DATE
CAD Support Associates, Inc.
DRAWING TITLE:
BUSHING MOUNT
SIZE
B
DRAWING NO.
DRAWING 9-3
REV.
SCALE:
DATE:
SHEET
OF

Index